FAIR WEATHER?
EQUITY CONCERNS IN CLIMATE CHANGE

Edited by
Ferenc L Tóth

Earthscan Publications Ltd, London

First published in the UK in 1999 by
Earthscan Publications Ltd

Copyright © Potsdam Institute for Climate Impact Research, 1999

All rights reserved

A catalogue record for this book is available from the British Library

ISBN: 1 85383 557 9 paperback
       1 85383 558 7 hardback

Typesetting by PCS Mapping & DTP, Newcastle upon Tyne
Printed and bound by Creative Print and Design Wales, Ebbw Vale
Cover design by Andrew Corbett

For a full list of publications please contact:

Earthscan Publications Ltd
120 Pentonville Road
London, N1 9JN, UK
Tel: +44 (0)171 278 0433
Fax: +44 (0)171 278 1142
Email: earthinfo@earthscan.co.uk
http://www.earthscan.co.uk

Earthscan is an editorially independent subsidiary of Kogan Page Ltd and publishes in association with WWF-UK and the International Institute for Environment and Development

This book is printed on elemental chlorine free paper

FAIR WEATHER?

Contents

List of Figures and Tables *vi*
Acronyms and Abbreviations *viii*
Foreword *x*

1 Fairness Concerns in Climate Change 1
  Ferenc L Tóth
2 Equity Issues and Integrated Assessment 11
  Steve Rayner, Elizabeth L Malone and Michael Thompson
3 Climate Change and Multiple Views of Fairness 44
  Joanne Linnerooth-Bayer
4 Empirical and Ethical Arguments in Climate Change Impact
  Valuation and Aggregation 65
  Richard S J Tol, Samuel Fankhauser and David W Pearce
5 Applying Fairness Criteria to the Allocation of Climate
  Protection Burdens: An Economic Perspective 80
  Carsten Helm
6 The Appropriateness of Economic Approaches to the Analysis
  of Burden-Sharing 94
  H Asbjørn Aaheim
7 Biases in Allocating Obligations for Climate Protection:
  Implications from Social Judgement Research in Psychology 112
  Volker Linneweber
8 Fairness and Local Environmental Concerns in Climate Policy 133
  Shuzo Nishioka
9 Justice, Equity and Efficiency in Climate Change: A Developing
  Country Perspective 145
  PR Shukla
10 Justice in the Greenhouse: Perspectives from International Law 160
  Frank Biermann
11 Equity in International Law 173
  Juliane Kokott
12 The Regulation of Greenhouse Gases: Does Fairness Matter? 193
  David G Victor

Addresses of Lead Contributors *207*
Index *208*

List of Figures and Tables

Figures

2.1	Local and global standpoints	14
2.2	Elements of the climate change problem	15
2.3	A tripolar policy space	19
2.4	A two-dimensional map of institutional discourse and human values	35
5.1	Equal distribution of burdens or benefits	84
6.1	Energy Technology Systems Analysis Programme (ETSAP) estimates of national costs of alternative emissions reductions in 2010	97
6.2	National costs of CO_2 emission reductions according to national top-down studies	99
6.3	Redistribution by cost effectiveness	100
6.4	Optimal emissions of CO_2 in cases A, B and C	103
6.5	Reductions in CO_2 emission by region at a 10 per cent reduction in world emissions according to alternative principles (percentage)	107
9.1	Carbon emissions per capita – 1993	151
9.2	Trajectories for convergence of per capita emissions	155
10.1	The norm square of the emerging legal concept of common concerns of humankind	169

Tables

2.1	Outcome fairness and asymmetry of losses and gains according to three ethical positions	26
2.2	Valid consent as framed by three ethical positions	27
2.3	Characteristics of three kinds of social solidarity as described in classic social science literature	31
2.4	Time, intergenerational responsibility and discount rates as framed by three kinds of solidarity	32
4.1	Annual monetized 2 x CO_2 damage in different world regions	67
4.2	Aggregate damages corrected for inequality (in billion US$)	72
4.3	Fankhauser's estimates for different positions on valuation and accountability	75
4.4	Implied inequality aversion (γ) as a function of the income elasticity of marginal utility (e) and empirical value ratio (V_r/V_p)	76

6.1	Assumptions about abatement costs and damage costs (per cent of GDP)	102
6.2	Selected equity principles and examples of related burden-sharing rules	105
9.1	Historic CO_2 and methane contribution by region, 1800–1988 (in percentages)	150
9.2	Per capita emissions (tonne of carbon per year) for IS92a, c and e	152
9.3	Cumulative carbon emissions for the period 1991 to 2100 for IS92 emission scenarios	153
9.4	Cumulative carbon emissions (1991–2100) for 'S' and 'WRE' emissions trajectories for stabilization of CO_2 concentrations	153

Acronyms and Abbreviations

ASEAN	Association of South-East Asian Nations
BaU	business as usual
CDM	clean development mechanism
CFC	chlorofluorocarbon
CO_2	carbon dioxide
COP3	third session of the Conference of the Parties (to the UNFCCC)
DENR	Department of Environment and Natural Resources (The Philippines)
ETSAP	Energy Technology Systems Analysis Programme
EU	European Union
GATT	General Agreement on Tariffs and Trade
GCM	global circulation model
GDP	gross domestic product
GEF	Global Environment Facility (World Bank)
GHG	greenhouse gas
GNP	gross national product
IA	integrated assessment
IAM	integrated assessment model
ICJ	International Court of Justice
ICLIPS	Integrated Assessment of Climate Protection Strategies
IGPB	International Geosphere–Biosphere Programme
IIASA	International Institute for Applied Systems Analysis
ILA	International Law Association
IPCC	Intergovernmental Panel on Climate Change
LDC	less developed country
LRTAP	long-range transboundary air pollution
MERGE	Model for Evaluating Regional and Global Effects (of GHG reduction policies)
NATO	North Atlantic Treaty Organization
NGO	non-governmental organization
NIE	newly industrialized economy
NIEO	New International Economic Order
NO_x	the sum of nitric oxide and nitrogen dioxide
NRC	US National Research Council
OECD	Organisation for Economic Co-operation and Development
PIK	Potsdam Institute for Climate Impact Research
ppmv	parts per million volume
RFF	Resources for the Future

SAR	Second Assessment Report (of the IPCC)
SIT	social identity theory
SO_2	sulphur dioxide
TWA	Tolerable Windows Approach
UKCCIRG	United Kingdom Climate Change Impacts Review Group
UN	United Nations
UNCBD	United Nations Convention on Biological Diversity
UNCED	United Nations Conference on Environment and Development
UNCLOS	United Nations Convention on the Law of the Sea
UN DPCSD	United Nations Department for Policy Coordination and Sustainable Development
UNDP	United Nations Development Programme
UNEP	United Nations Environment Programme
UNFCCC	United Nations Framework Convention on Climate Change
VOSL	value of a statistical life
WG	working group
WTA	willingness to accept
WTP	willingness to pay

FOREWORD

The Potsdam Institute for Climate Impact Research (PIK) is a Leibniz Institute, founded in 1992, with the Federal Ministry for Education, Science, Research and Technology and the Ministry for Science, Research and Culture of the State of Brandenburg each providing half of the funding. PIK currently has a staff of about 100 (1998), including approximately 75 scientists and guest scientists, as well as a number of students and temporary assistants. Further expansion is taking place at the institute site in the Albert Einstein Science Park in Potsdam.

The interdisciplinary nature of climate impact research, especially the interface between the natural scientific and socioeconomic dimensions of environmental research, is reflected at PIK in the close cooperation with partner institutes at national and international level. The flexible framework created for the institute enables new problems and issues to be taken up as they arise. As a centre of scientific innovation, PIK also coordinates international activities in the fields of climate impact research and Earth system analysis. The institute houses project offices for the International Geosphere–Biosphere Programme (IGBP) international research programmes, for example. Simulations of global change are performed on PIK's supercomputer using models and data drawn from various disciplines. The parallel computer (an IBM-SP2) boasts 20 gigaflops of computing power, making it one of the most powerful research computers in Germany.

One of the major projects at PIK is aimed at developing an Integrated Assessment of Climate Protection Strategies (ICLIPS).[1] The ICLIPS project is an international and interdisciplinary research activity involving a core team at PIK and six collaborating partner institutions. The ultimate question of managing the climate change problem is at what level should mankind stabilize climate with respect to anthropogenic interference. The project is developing an integrated assessment model (IAM) in order to help social actors explore the problem and the multifaceted implications of solution options by adopting the Tolerable Windows Approach (TWA). A typical TWA application involves the following steps:

- explicit normative definition of guardrails that exclude intolerable climate impacts as well as socioeconomically unacceptable mitigation measures;

---

1 Financial support to the ICLIPS project and the workshop on which this book is based has been provided by the German Federal Ministry for Education, Science, Research and Technology (Contract No 01 LK 9605/0) and by the German Federal Ministry for Environment, Nature Conservation and Nuclear Safety (Contract No 104 02 815). The views presented in the volume are those of the authors and do not necessarily represent the views of other ICLIPS partners or of the project's funders.

- integrated analysis of relevant and interconnected elements of the Earth system;
- determination of the set of all admissible climate protection strategies that are compatible with the predefined guardrails; and
- selection of specific policy paths by applying quantitative optimization models, by observing qualitative decision criteria or by seeking a compromise in a negotiation process.

The complex chain of causes and effects is described by the ICLIPS IAM. The model incorporates climate impacts, the climate system itself, relevant biogeochemical cycles, emissions of different greenhouse gases, allocation schemes of emission rights across nations, possible instruments for emission mitigation, and long-term dynamics of socioeconomic development.

IAMs, especially those framed in the cost-benefit paradigm, have often been criticized for not giving adequate attention to broader social dimensions of climate change. One of the most crucial points of criticism is related to the issues of fairness and equity in terms of differential vulnerability to, and potential impacts of, climate change as well as in terms of differential responsibility for the problem and ability to do something about it. While the importance of fairness and social considerations is generally recognized, there has been little effort to include them explicitly in IAMs. The ICLIPS project responded to this challenge by devoting an international and interdisciplinary workshop to the topic.

The Workshop on Fairness Issues in Climate Change brought together leading scientists in economics, sociology and social psychology, ethics, international law and political science to explore issues of fairness in the context of global climate change. There were two major objectives. The first one was analytical and tried to identify possible ways to address fairness issues in an integrated assessment framework. More specifically, how can one sort and structure the numerous equity concepts? What is the best possible advice one can get from the various disciplines? The second objective was related to making the assessment relevant for policy-making: what are the potentially useful ways to structure the various arguments, and what are the current and expected future analytical needs of the negotiations process?

We asked key participants to prepare presentations for the workshop and to cover the following items:

- a succinct overview of the state of the art of the fairness/equity concepts in the author's own discipline, reflecting recent developments in other disciplines;
- the most relevant lessons for the climate change problem;
- points to stimulate discussions among the various disciplines to identify common ground, potential complementarities, and mutually exclusive concepts;
- recommendations on how to include fairness/equity concerns in integrated assessments to improve both their scientific validity and policy relevance.

Chapters in this volume constitute papers prepared for and peer-reviewed and revised after the workshop. A notable characteristic of the material is that it looks at an important topic (fairness issues in the context of global warming) and addresses it from the perspectives of a broad range of social science disciplines. Taken together, these papers provide a concise picture of equity concepts in social sciences and assess the relevance of those concepts in the light of current policy processes in managing climate change.

In addition to the authors, many people contributed to producing this volume. Eva Hizsnyik coordinated the production process and served as technical editor. Thorough reviews and useful comments from a large number of referees are gratefully acknowledged.

Hans-Joachim Schellnhuber, Director
Ferenc L Tóth, Project Leader
Potsdam Institute for Climate Impact Research
December 1998

1 FAIRNESS CONCERNS IN CLIMATE CHANGE

Ferenc L Tóth

INTRODUCTION

The number of models that integrate the biogeophysical processes shaping the Earth's climate and the most relevant socioeconomic activities interfering with those processes have grown exponentially over the past decade. These models analyse the full cycle of anthropogenic emissions of greenhouse gases (GHGs), their concentrations in the atmosphere, the resulting climate forcing, and finally the impacts of the induced climatic change on society and economy. For economists and social scientists, the great advantage of this approach is that it provides a comprehensive framework for assessing the possible economic losses due to climate change (damage function) and for estimating the costs to slow or delay climate change (cost function). By creating the same metric for cost-and-benefit assessments, integrated assessment models (IAMs) can then be used to develop economically efficient climate policies. Several recent studies provide overviews and critical appraisals of IAMs (see Weyant et al, 1996; Rotmans and Dowlatabadi, 1998).

A frequent critical remark levelled at IAMs signifies the negligence of or, at most, marginal attention paid to equity concerns. The Potsdam Institute for Climate Impact Research (PIK) organized an international workshop to discuss underlying reasons and possible solutions. Based on presentations and deliberations at the workshop, chapters in this book aim to provide a multifaceted explanation of why this has been the case and how the problem could be alleviated. By looking at the concept of equity from the perspectives of a broad range of social sciences, the studies that follow are believed to contribute important pieces to the debate on fairness issues in the context of global climate change.

THE SCOPE OF THE BOOK

A useful and convenient starting point for our discussions is the chapter on equity prepared by Working Group III (WGIII) for the *Second Assessment Report* (SAR) of the Intergovernmental Panel on Climate Change (IPCC) (see

Banuri et al, 1996). The authors distinguish two separate categories of equity. The first involves procedural issues and attempts to sort out criteria for designing and implementing fair procedures for assessing climate-change related risks and for managing them. The two aspects of procedural equity involve participation in decision-making processes on the one hand and the process itself, 'most notably the principle of equal treatment before the law' (Banuri et al, 1996:85).

Although fair procedures are important preconditions to creating and operating an international regime in the first place, we leave it for practitioners to sort out problems of procedural equity and focus on the second category defined by the IPCC writing team. This category is related to the outcomes of decisions and is called consequential equity. Fairness issues in this category affect both the impact and the abatement side of the climate change problem. The bulk of the literature is devoted to the consequential aspects: justice or inequity in the hardship and/or blessing imposed by climate change, and just or unfair distribution of the burden associated with managing the problem, for example, reducing emissions.

In the consequentialist equity category, the global and long-term nature of the climate change problem raises two different bundles of equity issues. To a large extent, these can be considered independent of each other although some obvious relationships exist between them.

The global characteristic of the problem entails that action, harmful as well as protective, by any one will eventually affect everyone, but effects of the same global change will influence individuals, social groups, even whole countries very differently depending upon their geographical conditions, economic development and social institutions. These differences in vulnerability with respect to possible impacts, and in the ability to do something about the problem, are at the core of the debates on intragenerational equity.

Most GHGs emitted to the atmosphere modify the radiative balance of our planet and contribute to human induced changes in climate for many decades, even centuries. Any action by any given generation incurs costs for members of that generation while benefits accrue at least decades later and spread across many generations. This enormous time difference between costs and benefits is the source of what appears to be an eternal debate on intergenerational equity.

For recent contributions to the debate on intergenerational equity, we refer the reader to another chapter in the same IPCC volume (Arrow et al, 1996) and to the proceedings of a workshop that took place almost concurrently with ours (see RFF, 1999). An important precursor to both the IPCC chapter and the RFF conference was a set of papers presented at the Workshop on Integrative Assessment of Mitigation, Impacts, and Adaptation to Climate Change at the International Institute for Applied Systems Analysis (IIASA) in Laxenburg, Austria (see Lind, 1995; Manne, 1995; Schelling, 1995; and Tóth, 1995).

This book is devoted to the issue of intragenerational equity. The volume offers a grand tour across a broad range of social science disciplines in an attempt to explore what their paradigms and analytical frameworks can add to the overall debate on fairness in global change issues, especially climate change. Specifically, the collection offers relevant information for appraising to what extent equity concepts, as adopted in various disciplines, are compatible; are they complementary or mutually exclusive? As for the practical implications, we ask which earlier manifestations and practical implementations of those concepts have proven useful and successful in the past and which ones fall apart and remain captivating ideas with no practical relevance when exposed to the tough world of reality.

FAIRNESS CONCEPTS AND LOCAL EXPERIENCE

The starting point for our grand tour is a concise overview of various social science paradigms that lead to different analytical styles and produce diversely different contexts for thinking about, or ignoring, equity issues. Rayner et al (see Chapter 2) argue that there are deep and profound causes of what appears to be a relatively straightforward dichotomy between equity and efficiency, which often manifests itself as the contradiction between values and facts. The dichotomy apparently goes back to the sober difference between descriptive (empirically oriented) and interpretive (empirism cum explication) approaches, of which the former appears to dominate the contribution of social sciences to integrated assessments (IAs) of the climate problem. No wonder, then, that utilitarian studies (associated with the descriptive approaches) dominate over rights-based (stemming from the interpretive approaches) analyses in policy-oriented assessments of climate change.

Among the many informative deliberations in the chapter by Rayner et al, perhaps the most intriguing item is the two-dimensional map of institutional discourse and human values. The map offers an excellent framework on which to sort and locate standpoints and the (often hidden or implicit) values behind them. It also helps spot analytical tools and the (often overlooked) values involved in them. This approach to mapping moral landscapes is likely to be very useful in explicitly revealing one's stances regarding equity and may also help to identify cases when equity concerns are only employed to block agreement and/or block action.

While climate change and, more generally, global environmental problems are very different in nature from local pollution incidents, including those of getting rid of hazardous wastes, there still might be lessons to learn from decades of local environmental management cases. Linnerooth-Bayer (see Chapter 3) takes a closer look at how equity concerns factored in decisions on siting hazardous waste disposal facilities. Her treatise picks up many ideas related to ordering the huge diversity of equity concepts and their underlying paradigms surveyed in Chapter 2. She also presents some striking practical cases of public perception of, and opinion on, trade-offs between what appears to be economically efficient versus what are considered to be fair solutions of hazardous waste management. While it is clearly difficult, if at all possible, to work those conflicting value systems into the rational actors–utilitarian analytical frameworks, innovative techniques to consider them in decision analyses and in the policy process would be needed.

FAIRNESS IN ECONOMICS

There is widespread misconception about how and why economists tend to be preoccupied with efficiency and tend to leave aside issues of equity. It is too often disregarded that the large body of positive economics is built on observations and that theoretical constructs are tested against empirical data. Too often, results produced by these analytical techniques do not please those who dislike characteristics of the reality that go into constructing and calibrating economic models. Three chapters are devoted to economic aspects of the equity problems related to climate change.

By far the most controversial chapter in the report of WGIII of the IPCC SAR is the one trying to estimate social costs of climate change (Pearce et al, 1996). Tol et al (see Chapter 4) take a critical and self-reflective look at the criticism levelled at the various drafts and the final version of that chapter. The main issues include the choice between willingness to pay and willingness to accept; the difficulties involved in transferring benefits across individuals, social groups, and geographical regions; the use of equity weights in considering what is perceived to be unfair income distribution; and, perhaps the most disputed topic, the question of regionally differentiated versus uniform per-unit damage values. The main contribution of the chapter is a series of climate-change damage estimates performed on the basis of various alternative equity criteria proposed by those in disagreement with the welfare economic paradigm underlying the IPCC social cost estimates. Not surprisingly, the estimates based on the alternative assumptions vary to a great deal and tend to be significantly higher than traditional welfare economic indicators. It still remains a contentious issue to sort out which of those alternative propositions have any practical relevance and which ones should be discarded as unrealistic or outright utopian.

Some authors argue that due to the equity issues involved in the climate change problem, perceived current distortions in the global distribution of wealth should be taken into account and thus climate policy should be linked to development policies. In contrast, Helm (see Chapter 5) presents an analysis of fair burden-sharing of greenhouse gas reduction measures by separating the climate issues from the frantic dispute area of global welfare distribution. This is a rather practical analytical strategy with possibly far-reaching implications for policy. The bottom line is: seek solution to the climate change problem and do not try to salvage the world at the same time. An equitable solution to the climate change problem should be possible without detrimentally affecting the overall global distribution of wealth. In his chapter, Helm proposes five simple criteria that should be fulfilled in order to make the climate related burden-sharing a fair one. Taken together, these widely acceptable, largely common sense criteria define a complex system of burden allocation mechanism. Although details of its practical implementation are far from obvious, the scheme appears to be compatible with instruments currently discussed in policy fora.

An evergreen subject in the economic analysis of social policy is the relationship between efficiency and equity both in the theoretical foundations of the analysis as well as in the outcome. While analytical tools and models, including those that are incorporated in integrated assessment models of the climate change problem, properly handle the efficiency aspects, they still fall short of addressing equity concerns properly.

In Chapter 6, Aaheim takes this dilemma as a starting point and provides a concise appraisal of the role of economic analysis in supporting social choices on climate policy, with a special focus on integrating fairness concerns. In taking stock of what has already been a useful contribution, the author refers to the robust analytical frameworks to address various issues of distribution adopted in order to examine the highly contentious problem of burden-sharing. Aaheim also explores some unutilized potential of already available and well-established analytical tools. A detailed assessment of the widely diverging climate change related cost-benefit ratios of individual nations would not only provide an improved understanding of the positions that countries are taking at international negotiations but might also contribute to finding fair and generally acceptable agreements by providing some basic numbers for side-payments or other forms of compensation.

Finally, Aaheim discusses largely unresolved issues that require bridging the gap between moral philosophy and economics. The central question here remains how to use welfare functions at the international level where there is no authority to ensure that the welfare function reflects the priorities of the public. Since different assumptions about the income distribution among nations imply different welfare functions, the resulting spread in national damage estimates makes national cost-benefit ratios rather hazy indicators as well.

FAIRNESS IN SOCIAL SCIENCES

From the cruel world of economics we then proceed to the colourful world of psychology and sociology. An increasingly popular proposal has been emerging over the past few decades that seeks to find the dominant explanations of human behaviour in 'culture', a shorthand for values, aspirations and customs that characterize civilizations in different regions of the world (see, for example, Sowell, 1994; Huntington, 1996). In contrast, many scholars maintain that pure economic interests, political ideas and actual policies provide better explanations of what happened in the past and what is likely to happen in the future. While the debate is bound to continue, it is important to observe the increasing recognition of the potential importance of culture-based explanations in understanding regional differences in perceiving and interpreting global environmental risks and their local repercussions. The following three chapters are devoted to this issue.

Linneweber (see Chapter 7) begins by providing a concise overview of the key concepts of justice and the main contexts in which fairness issues are investigated in studies of social psychology, ranging from individuals through micro-level (interpersonal relationships) and meso-scale (institutions and organizations) social systems all the way to intergroup relations. The author then explores the notion of environment as it is defined and interpreted at these different levels and the variety of ways in which the biophysical environment is used to explain relationships among individuals and groups within and across those levels. The culprit–victim relationship appears to be particularly relevant for global environmental issues such as climate change where the diversity of past and present sources of emissions, and the multitude of present and future impacts, offer a fertile soil for quarrelling about responsibility and restitution.

Linneweber also discusses the diversity of values and aspirations even within relatively homogeneous societies and the difficulties involved in consolidating them. These problems become next to incommensurable when we need to reach cooperation and agreement across a broad range of culturally diverse societies. The next two chapters provide valuable insights into these issues.

Nishioka (see Chapter 8) draws attention to a largely neglected aspect of global warming. Climate change will manifest itself in the form of myriads of local impacts such as implementing climate policies, which will also involve scores of local mitigation actions. Responses to local impacts and prospects for local emission-reduction endeavours do not only depend on geography, climate, resource endowment and other physical conditions. They are largely determined by economic, technological and social factors, among which indigenous culture and tradition are particularly important. The author explores these ideas by linking local ecological consciousness to climate policy in the Asia–Pacific region. He identifies cultural values and elements of traditional wisdom which could and should be mobilized not only for successful local implementation of global climate policies but for promoting the transition to an environmentally sustainable development path as well.

It is by no means a new statement that developing country perspectives on the climate change problem and its management strategies differ considerably from those of industrial countries. Shukla (see Chapter 9) explores the equity aspects of those differences both at the conceptual level as well as their practical implications for international negotiations. While industrial countries tend to emphasize the importance of economic efficiency in designing an international regime to address climate change, developing countries are concerned with how the mitigation burden should be distributed among nations. The latter implies a mitigation regime that minimizes welfare losses, which is unlikely to put much load on developing countries since their per capita incomes and emissions are expected to remain considerably lower than those of industrial countries. An efficient regime, nevertheless, would require developing countries' participation from early on. According to Shukla's conclusion, a prerequisite to this is to make fairness issues an essential item on the post-Kyoto negotiations agenda.

Perspectives from Law and Political Science

It is one question what moral philosophers and other analysts have proposed for getting things right among nations. It is quite another what has been actually codified in international law to regulate relationships among nations. Historically, the answer is surprisingly little, which seems to be the shared conclusion of the next two chapters; but this appears to be changing very quickly.

One possible starting point, according to Biermann (see Chapter 10), is what the majority of governments considers equitable. National governments are the entities establishing positive international environmental law. Legal concepts of justice for the case of global climate change can thus be derived from the general principles already embodied in existing international environmental legislation. In this context, the author provides useful insights into the relationship between empirical state practice and normative concepts of fairness in international law.

Recent global environmental legislation (eg, the United Nations Framework Convention on Climate Change (UNFCCC, 1992) and the United

Nations Convention on Biological Diversity (UNCBD, 1992) introduced the term 'common concern of humankind'. From a legal perspective, this concept is related to a sophisticated set of principles such as solidarity, equal participation and differentiation. Biermann suggests that this assembly of principles could form a solid foundation for operationalizing the concept of justice in international environmental law.

It is worth noting that the set of principles discussed by Biermann is also congruent with a key concept in economics – ability to pay – which is particularly relevant in global environmental affairs. This is especially important because implementing and enforcing of international environmental regulations might inevitably involve economic sanctions which will need to be legally harmonized with international economic agreements. Harmonizing basic principles underlying these regimes is indispensable for their effective operation.

The second treatise on international legal aspects of fairness in climate change policies in this volume by Juliane Kokott (see Chapter 11) provides a systematic overview of the equity issue, starting from the broad field of general international law. The author presents historical origins and the evolution of equity principles both at the conceptual level and by quoting specific cases and rulings of the International Court of Justice. Converging with international environmental law, the intricacies of applying legal manifestations of what appear to be simple equity principles in practice are also illustrated by a series of cases. Armed with lessons from analysing broad principles and earlier cases, Kokott then presents a thorough legal analysis of equity in existing international legislation on climate change. She identifies virtues as well as potential pitfalls for implementation.

Kokott's chapter confronts implementation issues from the legal perspectives. But how much does fairness matter in reality? The summary of the IPCC chapter quoted earlier poses two strong assumptions: 'countries are unlikely to participate fully unless they perceive the arrangements to be equitable' and 'governments will find it easier to comply with international obligations if their citizens feel that the obligations and benefits of compliance are distributed equitably' (see Banuri et al, 1995:83). Surprisingly enough, the chapter does not provide much evidence in support of these assumptions, perhaps because they are widely considered as conventional wisdom.

Looking closer, though, there appears to be little correlation between how fairness principles were built into the agreement and how successful the agreement was. In some cases, differentiation might even weaken implementation. The Montreal Protocol is often praised for observing fairness concerns. In reality, the grace periods given to less developed and formal centrally planned countries have originated perverse developments. Smuggling ozone-damaging chlorofluorocarbon-12 and similar chemicals (CFCs) from countries where their production is still legal to the United States and Western Europe where their production is now banned has become a multi-billion dollar illegal business. This leakage attenuates the environmental objective of the treaty which was fast and drastic reduction of emissions of ozone-depleting substances. Moreover, smuggling increases the costs of compliance because the price of CFC substitutes declines more slowly than originally envisioned under full compliance. As new production capacities of those substitutes benefit from economies of scale with significant delays, the full price will be even higher. It might be most instructive for future international agreements to ask the question: how do the leakage costs compare with those of an attractive financial compensation mechanism or technology transfer programme that would have made developing countries more actively comply with the agreement from the very beginning?

Victor (see Chapter 12) takes a close look at fairness–compliance assumptions in the light of lessons learned about the implementation and effectiveness of numerous earlier international environmental agreements. His conclusion is that equity concerns matter little in the success of negotiating and implementing those agreements. Even in cases where fairness seems to play some role, it is instead willingness to pay – a correlated factor – at work. The author argues that if parties to an international environmental agreement take the trouble of deviating from the simplest across-the-board commitments, there are many other criteria to be considered in negotiating differentiated commitments. Fairness might be one criterion but probably not the most important.

In his essay, Victor draws on an extensive analysis of factors that have been fundamental to successfully implementing international environmental agreements. Reasons behind the very few cases in which differentiated commitments were agreed upon are diverse. Two major factors are proposed by the author that stand out in importance: public perception of the possible costs related to climate change that shapes the willingness-to-pay component, on the one hand, and the status of democratic decision-making that prevents or facilitates the articulation of those perceptions in the policy-making process, on the other.

CLOSING REMARKS

Equity is an intriguing aspect of the climate change issue. One can witness a strange mixture of well-taken and fully justified arguments about historical responsibility, current ability to pay for climate policies, and future vulnerability and long-term benefits of emission abatement. Yet, equity principles are often used as a veil to hide the real interests associated with greenhouse-gas related resources or production capacities and to avoid or delay action. Attempts to manage the climate change problem often appear to be a battle field of conflicting interests. Fairness concerns offer almost unlimited possibilities to support diverse camps.

Many different fairness criteria have been proposed in the political process and analysed in scientific literature. They have profoundly different consequences in terms of who should do what, at whose cost, and when to alleviate the climate problem. One of the current stumbling blocks of the political process can also be traced back to basic equity issues. Developing countries (listed as Non-Annex I countries in the UNFCCC) are unwilling to embark upon any commitment and use largely equity-based arguments: industrialized countries are mainly responsible for past emissions; they can afford to pay the costs; less developed countries (LDCs) need to maintain their rights to develop and this excludes any share in the climate mitigation burden now and in the foreseeable future. Industrial countries (listed as Annex I countries), on the other hand, argue that globally significant and hence meaningful reduction is unimaginable without the effective participation of Non-Annex I countries, even if the lion's share of the costs is paid by industrial countries.

Integrated assessment models have provided valuable insights into many aspects of climate change. Over time, these models should also be able to contribute to the equity debate and help to work out viable solutions.

Each station in our grand tour across social sciences has provided something important and relevant on fairness. Authors also proposed ways of incorporating these perspectives in integrated analyses. The basic problem, nevertheless, remains unresolved: how to formalize these contributions so that they can be part of an integrated assessment modelling framework, preferably on an equal footing with other components that have a longer tradition in formal modelling. This clearly does not solve this problem. But it is hoped that the material collected here will help future efforts in fulfilling this challenging task.

REFERENCES

Arrow, K J, Cline, W R, Maler, K-G, Munasinghe, M, Squitieri, R and Stiglitz, J E, 1996, Intertemporal equity, discounting, and economic efficiency, pp 125–144 in: Intergovernmental Panel on Climate Change (IPCC), *Climate Change 1995: Economic and Social Dimensions of Climate Change*, Contribution of Working Group III to the Second Assessment Report of the IPCC, Cambridge University Press, Cambridge, UK.

Banuri, T, Göran-Mäler, K, Grubb, M, Jacobson, H K and Yamin, F, 1996, Equity and social considerations, pp 79–124 in: Intergovernmental Panel on Climate Change (IPCC), *Climate Change 1995: Economic and Social Dimensions of Climate Change*, Contribution of Working Group III to the Second Assessment Report of the IPCC, Cambridge University Press, Cambridge, UK.

Huntington, S, 1996, *The Clash of Civilisation and the Remaking of World Order*, Simon & Schuster, New York.

Lind, R C, 1995, Intergenerational equity, discounting, and the role of cost-benefit analysis in evaluating global climate policy, *Energy Policy* 23(4/5):379–389.

Manne, A S, 1995, The rate of time preference: implications for the greenhouse debate, *Energy Policy* 23(4/5):391–394.

Pearce, D W, Cline, W R, Achanta, A N, Fankhauser, S, Pachauri, R K, Tol, R S J and Vellinga, P, 1996, The social costs of climate change, pp 179–224 in: Intergovernmental Panel on Climate Change (IPCC), *Climate Change 1995: Economic and Social Dimensions of Climate Change*, Contribution of Working Group III to the Second Assessment Report of the IPCC, Cambridge University Press, Cambridge, UK.

RFF (Resources for the Future), 1999, Proceedings of the Workshop on 'Discounting in Intergenerational Decision Making', RFF, Washington, DC, forthcoming.

Rotmans, J and H Dowlatabadi, H, 1998, Integrated assessment modelling, pp 291–377 in: S Rayner and E L Malone, eds, *Human Choice and Climate Change, Vol 3: The Tools for Policy Analysis*, Battelle Press, Columbus, Ohio.

Schelling, T C, 1995, Intergenerational discounting, *Energy Policy* 23(4/5):395–401.

Sowell, T, 1994, *Race and Culture: A World View*, Basic Books, New York.

Tóth, F L, 1995, Discounting in integrated assessment models, *Energy Policy* 23(4/5):403–409.

UNCBD, 1992, *United Nations Convention on Biological Diversity*, United Nations, New York.

UNFCCC, 1992, *United Nations Framework Convention on Climate Change*, United Nations, New York.

Weyant, J (lead author), 1996, Integrated assessment of climate change: An overview and comparison of approaches and results, pp 367–396 in: Intergovernmental Panel on Climate Change (IPCC), *Climate Change 1995: Economic and Social Dimensions of Climate Change*, Contribution of Working Group III to the Second Assessment Report of the IPCC, Cambridge University Press, Cambridge, UK.

2   EQUITY ISSUES AND INTEGRATED ASSESSMENT

Steve Rayner, Elizabeth L Malone and Michael Thompson

INTRODUCTION

The question posed by the organizers of this workshop was essentially whether and how fairness and equity issues can and should be incorporated into integrated assessments of climate change that are designed to inform policy and other decision-makers about the potential outcomes of both climate change itself and the policies designed to deal with it. To the best of our knowledge, no one has asked the question whether and how efficiency issues can and should be incorporated into integrated assessment. After all, the assumption that decision-makers are responsive to prices underlies the economic modules of all integrated assessment models. Efficiency, unlike equity, is inherent in the very idea of integrated assessment.

The fact that we view the issue of incorporating equity considerations in integrated assessment as problematic, while building critical components of our models on unquestioned assumptions about efficiency, points to a pervasive feature of Western intellectual life since the Enlightenment – that is, the dichotomy of facts and values. Efficiency can be measured in direct physical terms, such as joules, and indirect quantities, such as dollars. Efficiency, therefore, belongs to the domain of facts (or, rather, it does so long as we remain unaware of the particular idea of fairness that is embedded in the very notion of efficiency). However, we have yet to devise direct measures of fairness and are frequently bothered by indirect monetary yardsticks.

The dichotomy of facts and values is enshrined in recommended scientific procedures for risk analysis. For instance, the US National Academy of Sciences (NRC, 1982) has long advocated a three-stage approach: establish the probability and magnitude of loss (a technical process); evaluate the benefits and costs (a social process); and manage the resources accordingly (a hybrid process). For just as long, social scientists have argued against this procedure for at least two reasons:

- Selection of issues for the attention of technical analysis (for example, nuclear reactors or hydropower dams) already embodies value commitments.

- Selection of methods of analysis (for example, linear or nonlinear dose response curves) already embodies value commitments.

The technical reduction of risk to probabilities and magnitudes of costs fails to reflect the cluster of concerns about trust, liability and consent to activities that are inherent components of the concept of risk in everyday usage.

These critical social scientists have instead argued for a holistic perspective on risk analysis which dispenses with the dichotomy of facts and values (Krimsky and Golding, 1992). In recent years, the holistic perspective has begun to change the practice of risk analysis on the cutting edge (NRC, 1996). However, it has had very limited impact on integrated assessment of climate change, which is essentially risk analysis on a global scale. This is because the dichotomy of facts and values is so deeply rooted in our Enlightenment heritage and fundamentally shapes the research methods and explanatory frameworks of the social sciences and the social practices of politicians. Before we can expect to reconcile facts and values (as well as efficiency and equity) in integrated assessment work, we need to understand something of the origins of the intellectual division of labour introduced by the Enlightenment. This chapter first looks at several interpretive models that help to define and explain the current equity debates about responses to climate change. It then discusses equity issues in climate discourses, distributional principles and allocational issues, procedural fairness, and the relationship of equity issues to social solidarity. A two-dimensional map of human values provides a tool for bringing equity issues into integrated assessment. The chapter closes with suggestions for ways to incorporate equity concerns into integrated assessment.

Two Styles of Intellectual Labour: two Kinds of Social Science

The characteristic ways of thinking within the natural sciences and the humanities have diverged over the course of the last three centuries since the Enlightenment. The internal coherence of each has been enhanced by insulation from the other. In contrast, Rayner and Malone argue (1998) that the social sciences have led a less comfortable existence, growing up astride the fault where the tension between knowledge and experience is most acute. Some social scientists have explicitly taken their terminology and ways of conceptualizing from natural science paradigms; others have explored instead the social scientific implications of history, philosophy, theology and other fields within the humanities.

The social sciences, therefore, are characterized by the uneasy coexistence of two distinctive approaches of the natural sciences and the humanities: the *descriptive* paradigm and the *interpretive* paradigm (Rayner and Malone, 1998). The term descriptive here refers to research-based descriptions of social systems rooted in natural science metaphors – for example, in terms of mass balances, thermodynamics or stocks and flows. The interpretive approach refers to the analysis of the values, meaning and motivation of human agents. The descriptive approach derives from empirical, Newtonian science; the interpretive from classical Greek philosophy through Kant's assertion that knowledge is derived not only empirically, but also from the mind itself.

The advantages of the descriptive approach include drawing on the legitimacy of the natural sciences by mirroring their methods and paradigms, laying out large problems and rendering them more tractable, comparing the efficiency of candidate policies, and providing a basis for control of tangibles (for example, counting tonnes of coal to limit the number that can be burned). These advantages of descriptive methods rest to a large extent on their simplifying assumptions about complex human behaviours. The descriptive approach is therefore capable of high levels of aggregation and large-scale, global analyses.

The interpretive approach emphasizes variation and complexity in human motivation. One of the primary foci of the interpretive method is the understanding of how humans draw meanings out of the data presented to them, emphasizing the sociocultural and institutional (including political) processes involved. The systematic study of such processes yields understanding that can help policy-makers make critical choices, knowing what assumptions and decision elements underlie those choices. The interpretive approach, therefore, tends to produce highly disaggregated, local-level analyses.

Linking the local and the global is frequently cited as one of the most challenging aspects of climate change and other global changes. Few attempts have been made to validate studies at one end of the scale with those at the other, and research studies at the middle levels are sparse, especially those that articulate connections among individuals, groups of various sizes and cultural backgrounds, and global-level players in climate issues. Throughout the social sciences there is poor understanding of how to articulate human behaviour at the local level to the behaviour of the global social and economic system. This is an important observation for our symposium, because local decisions tend to be more susceptible to equity considerations than global debates in which efficiency predominates (Elster, 1992).

Linking the local and the global cannot be achieved simply by increasing the scale and quantifiability of interpretive analysis to meet a more thickly textured descriptive analysis as it attempts to accommodate lower levels of aggregation. The gap between the two approaches is not merely spatial but raises fundamental issues of what kinds and sources of knowledge we value as analysts. Contrasting 16th century iconography with contemporary satellite photographs, Ingold (1993) illustrates the situation by tracing the change in the standpoint of human inquiry from the Enlightenment to the present (see Figure 2.1). In Maffei's *Scala Naturale* of 1564, the scholar is shown at the centre of the environment consisting of 14 concentric circles forming a giant stairway, the ascent of which, step by step, affords more comprehensive knowledge of the world through experience within it. In modern satellite imagery, the scholar experiencing the world from within is displaced by an observer viewing the world from without.

Style, scale and implications for equity considerations in integrated assessment

Social science collaboration in integrated assessment research programmes that address climate change exemplifies the descriptive approach and emphasizes its compatibility with natural science perspectives. This work has focused

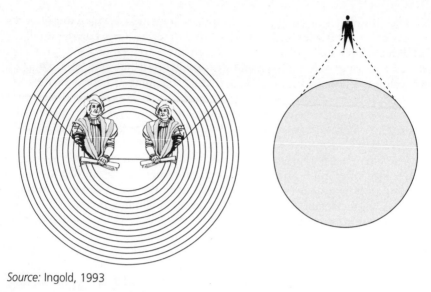

Source: Ingold, 1993

Figure 2.1 Local and global standpoints

on extending the framework already established by the natural sciences. The framework consists of a four-box conceptual model: quantified emissions of greenhouse gases, atmospheric chemistry, climate and sea level, and ecosystems. Within this framework (see Figure 2.2), the social sciences provide highly aggregated data on human activities leading to greenhouse gas emissions. These data can be used to drive natural science models of global atmospheric chemistry and physics. In turn, the natural sciences aim to model climatic impacts on the managed and unmanaged ecosystems upon which humans depend. At this point, social scientists are invited to project the outcomes of these changes for large-scale patterns of agricultural and industrial activity, stimulating macroeconomic and technological responses which, in turn, may eventually alter anthropogenic emissions estimates. The outputs of such research are presented as data: grist to the decision-maker's mill.

The need for large-scale analyses, the origination of studies in the natural sciences and the need for quantified data to run computer-based models are factors that help to explain the predominance of descriptive research on climate change issues. However, at the human level, the issue of climate change is ultimately an ethical issue, that is, one firmly rooted in the province of the interpretive paradigm. Just as technical analyses and arguments prove insufficient to persuade communities to use imported technologies, human decisions about the threat of climate change are not merely technical. They are decisions about equity, what is fair in our relationships with each other, and about natural ethics, what is right with respect to our relationships with nature.

While the analysis and description of ethical systems as social phenomena belong in the social sciences, the normative exercise of ethical reasoning is traditionally the province of philosophy and theology. Therefore, bridging the gap between the descriptive and interpretive traditions in social science – and even exploiting the commonalities between interpretive approaches in the social sciences and the humanities – is an imperative for effectively linking the practice of scientific and moral reasoning in confronting global climate change.

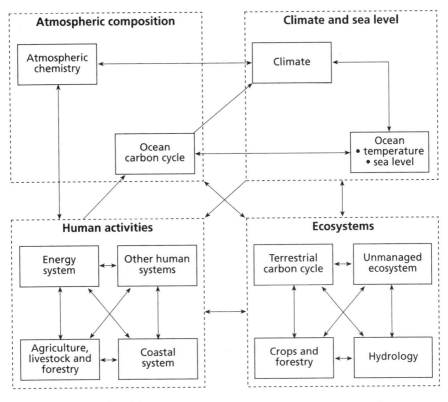

Source: Watson et al, 1996

Figure 2.2 Elements of the climate change problem

A further practical imperative is to understand human choice in social change as well as aggregating market behaviour and mapping demographic change. Following in Ricardo's footsteps (Gudeman, 1988), the descriptive method assumes continuity of human preferences and consistency of human behaviour over time. The descriptive models do not seek to explain what humans value, and therefore cannot anticipate sudden ruptures in social behaviour resulting from changes in values and the institutional arrangements that embody those values, or assess the potential for changing human motivations through political or other forms of social intervention. Because the vast array of human behaviour involved is effectively compressed into drivers and responses, the descriptive approach encourages the adoption of descriptive kinds of explanations for the actions of social systems and the individuals within them. Hence, policy prescriptions based solely on the descriptive paradigm are limited to instrumental tinkering with technology and prices.

However, social systems involve *human choice*; unlike a billiard ball shot across a table or an electron orbiting the nucleus of an atom, a human being has the ability to make conscious decisions about the directions in which he or she is moving and at what speed. Although social decision-makers are often constrained by their own paradigms and the initial conditions of the problem they are attempting to resolve, the existence of human choice means that their actions cannot be accounted for in purely deterministic terms.

Interpretive social science also brings to the conduct of scientific inquiry an awareness of the impact of human choice by introducing *reflexivity* into the research and policy-making processes. Reflexivity is the self-conscious examination of the implicit assumptions that are inevitably embedded in any analytical approach. In the descriptive approach, assumptions may be systematically laid out but are typically unquestioned within the research study. This is consistent with the outside observer stance of the descriptive researcher. If, however, these assumptions are made explicit, the researcher has an opportunity to question them, rather than take them as given. The interpretive contribution of keeping assumptions visible adds meaning to the research results by providing clear boundaries and caveats. Therefore, if research in the two paradigms can be integrated, a more complete analysis is accomplished.

Combining the reflexivity of the interpretive with the instrumental knowledge of the descriptive method seems an obvious course for improving our understanding of the human dimensions of climate change and its use in policy-making. A reflexive social science which draws effectively on both the descriptive and interpretive methods could help identify the multiplicity of characteristics needed to achieve a successful solution and the political problems, both immediate and longer term, associated with varying alternative climate management strategies.

Utility and rights-based approaches to decision-making

The descriptive and interpretive approaches respectively embody quite different normative imperatives for decision-makers. The tendency of the descriptive paradigm towards high levels of aggregation gives rise to a top-down decision-making rationality. The quantitative aspect of the paradigm leads to an essentially utilitarian perspective on decision-making. That is, the practice of inventorying the stocks and flows of goods and bads creates the conditions for a decision framework based on a technique for calculating societal happiness measured by the distribution of goods and tools such as the utilitarian calculus of Jeremy Bentham (1748–1832). It is but a short step from calculating what would contribute to the greatest happiness of the greatest number to creating the imperative to pursue that goal. Hence, efficiency is not merely a technical issue or an indication of rational behaviour within utilitarianism, but is also an intrinsically moral imperative that arises from the descriptive paradigm itself.

The imperative to provide for societal good at the highest level of aggregation provides no guidance for securing the happiness of minorities and individuals, or even of those individuals in the happy majority.

> *The guiding criterion for policy is the greatest good for society, quantitatively defined. But contemporary utilitarians, primarily economists and theorists of public choice, like Bentham, still have no principle for distributing this social good according to manifest principles of equity.*
>
> Heineman et al, 1990:40

In contrast, the tendency of the interpretive tradition to focus on the individual rather than the nation state directs the attention of scholars working in that paradigm to the particular circumstances of decision-making, rather than to the aggregate outcome. Increasing insight into the diversity of motives, values and preferences of individuals actually tends to frustrate utilitarian social accountancy, which depends on blending out such distinctions in the process of aggregation. It is hardly surprising, therefore, that the insights of the interpretive paradigm are not merely considered irrelevant to, but actually have to be excluded from, utilitarian decision-making in order to preserve the rationality and legitimacy of the utility principle.

As a result, the distinction between two social science paradigms lies at the heart of the crisis of governance that pervades the local, national and global communities at the close of the 20th century – that is, the tension between interdependence and independence, between pursuit of the greatest happiness of the greatest number, and the assertion of individual, local or ethnic rights that ought not to be violated even at the expense of the aggregate good. Whereas Kant's principle that every person is to be regarded as an end in him or herself is generally recognized as a form of the doctrine of human rights, Bentham dismissed the concept of rights as 'plain nonsense' and the imprescriptible rights of man as 'nonsense on stilts' (Russell, 1946:742). He denounced the articles of the Déclaration des Droits de l'Homme as falling into three classes: 'those that are unintelligible, those that are false and those that are both' (Russell, 1946:742).

Similar vituperation for the social inefficiency of rights-based ethics is not unheard of among contemporary utilitarians. For example, in response to proposals by sociologist Robert Bullard that current inequities in the distribution of environmental burdens on minorities and the poor should be addressed on an environmental-rights basis, rather than according to risk-based criteria, economist Albert L Nichols responded:

> *This framework has considerable popular appeal, but* it ultimately is counterproductive from the perspectives of both society as a whole and even the specific groups it tries to champion. *Moreover, it provides little practical guidance to environmental decision makers trying to set priorities...Bullard's proposed environmental justice framework makes continued inequities in protection more likely.*
>
> Nichols, 1994:267

Bullard replied that his proposals:

> *...are no more regressive than the initiatives taken in the 19th century in eliminating slavery and 'Jim Crow' measures in the United States. This argument was a sound one in the 1860s when the 13th Amendment of the Constitution was passed despite the opposition of proslavery advocates, who posited that the new law would create unemployment (slaves had a zero unemployment rate), drive up wages (slaves worked for free) and inflict undue hardship on the plantation economy (loss of absolute control of privately owned human property).*
>
> Bullard, 1994:260

Clearly these are not merely technical arguments about the best way to clean up the environment. Similar clashes between the utilitarian and rights-based views arise over the projected costs of climate change. In response to damage estimates that climate change will result in a decline in global productivity of less than 1 per cent over the course of the next century, utilitarians have expressed the view that only low-cost mitigation measures can be justified. On the other hand, those who espouse a rights-based approach point out that even less than 1 per cent of global productivity over 100 hundred years may translate into considerable suffering and premature death for millions of poor people in vulnerable regions of the less industrialized world.

Typically, policy-makers do approach risk characterization as a technical issue, such as exposure pathway, dose and response. However, the technical information is unlikely to influence citizens living in an area where environmental risk is present. If the science–policy relationship remains squarely in the descriptive mindset, the response to citizen objections is likely to be collecting and disseminating more technical data. The inclusion of values and worldviews will not be an option. However, if the science–policy relationship is collaborative and includes interpretive data, science will be 'more integrated into the policy context, more contextual and openly value laden, less oriented to mastery over natural and social processes and more accessible to the public at large' (Robinson, 1992:249). To the extent that scientists, policy-makers and the public can learn about each other's positions and preferences, the solution space for integrated assessment becomes larger and policy solutions may become more plausible.

Interpretive science in climate change

Several interpretive models are available to analyse equity issues and contribute to sound decision-making. The differences in people's views of nature, for example, provide an explanation for the varied (and opposed) rationales for diverse response actions (see Nash, 1967). Different views of nature have been classified as *catastrophist* and *cornucopian* (Meadows et al, 1972; Cotgrove, 1982; Bloomfield, 1986). Others have been identified with the market–hierarchy dichotomy. Three sets of general environmental models (Kempton et al, 1995) include nature as a limited resource upon which humans rely; as balanced, interdependent and unpredictable; and as a cultural view of society and nature. Kempton et al's findings (1995) closely match the *myths of nature* identified by Holling (1986): nature as benign, ephemeral or perverse/tolerant, to which Thompson (1987) added a fourth myth, that of nature as capricious.

Following Rayner (1991), the climate-change policy debate can be described as constituting a tripolar policy space defined by competing institutional voices, each representing a different policy bias (see Figure 2.3). The first is the activist voice that identifies *profligacy* in resource use as the underlying problem. The second voice blames economic inefficiency resulting from incorrect *pricing* or allocation of *property rights* for energy and forest resources. This voice is raised in disagreement with the first voice and the third voice, which diagnoses rapid *population* growth as the principal culprit behind climate change. As well as asserting its diagnosis of the factors under-

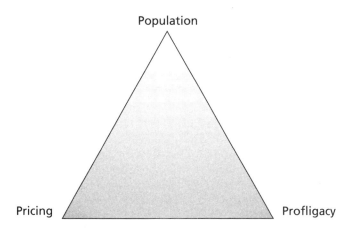

Source: Thompson and Rayner, 1998

Figure 2.3 A tripolar policy space

lying and driving climate change, each voice offers its own solutions – frugality, price reform and population control, respectively (Thompson and Rayner, 1998; Rayner and Malone, 1998).

Each voice can be heard to some degree in almost all institutions and at all levels from household decision-making to international negotiations. Approaching these debates at the level of discourse, rather than through the classifications conventionally used in policy analysis (the state, national governments, pressure groups, etc), captures the idea that institutions or even individuals are not unitary actors that can unambiguously represent (still less implement) a particular policy. Voice should not be equated with person. The speaker who uses only one voice is likely to be marginalized. Single individuals may use several voices. Indeed, speakers must adjust their rhetoric if they are to be fully heard within any particular institutional framework. As we describe in more detail later in this chapter, an institution is most often a pluralistic system in which competing and contradictory interests jockey for position.

Conclusions that require us to decide who is right miss the point. We need to turn ourselves firmly in an opposite, and much more reflexive, direction. What this analysis of climate change discourses tells us is:

- These are contradictory certitudes – they cannot be reconciled because each position actively defines itself in contradistinction to the others.
- Issues of diagnosis and prescription are compounded when viewed through the lens of equity.

EQUITY ISSUES IN CLIMATE DISCOURSES

Equity issues entered the climate change discourse in the late 1980s, arising out of the realization that certain nations or regions may actually gain from the impacts of change while others will be disadvantaged (Kasperson and Dow, 1991). For example, growing seasons in high latitudes might be extended while lower latitudes experience drought. Ports previously icebound throughout the winter months might become accessible year round, while major navigable rivers elsewhere could be closed to traffic during the summer because of reduced stream flow. At that time there was relatively little interest in climate change among the less industrialized tropical nations, whose elites regarded it as a problem created by, and affecting, the wealthier nations of temperate latitudes.

However, the discussion was soon extended beyond the equity implications of impacts themselves to the actions that countries could or should take to prevent climate change or to adapt to its consequences. The implications of measures to reduce greenhouse gas emissions on the development trajectories of the less industrialized nations became a prominent concern, leading government representatives to focus on the historical responsibility of industrialized countries for greenhouse gas emissions and on the current disparity in per capita emissions between North and South. At the same time, rapidly rising trends in Southern emissions caught the attention of Northern analysts, who emphasized to their policy-makers that their own efforts to control emissions would be fruitless without the participation of nations undergoing economic and technological development (Fulkerson et al, 1989). Southern interest in climate change increased drastically with the opportunity to make cooperation conditional upon Northern economic and technological support for their development aspirations.

As a result, the language employed in the first principle in Article 3 of the 1992 Framework Convention on Climate Change (FCCC) represents a classic diplomatic formulation that permits all parties to enjoin equity – that is, fairness or justice – while retaining quite different notions of what would actually be fair and what duties, obligations or commitments fairness would entail (UN, 1992). The leadership duties of industrialized country parties are left similarly vague. The most recent agreement, in Kyoto (UN, 1997), continues to carefully word its points while leaving details to be settled later. They will probably have to remain so with respect to guiding principles about which, this chapter will argue, it may be more difficult to obtain agreement than to negotiate practical measures.

Henry Shue (1994) breaks the issue of fairness in climate change policy into four questions:

- What is a fair allocation of the costs of preventing the global warming that is still avoidable?
- What is a fair allocation of the costs of coping with the social consequences of the global warming that will not, in fact, be avoided?
- What background allocation of wealth would allow international bargaining (about the first two points) to be a fair process?
- What is a fair allocation of emissions of greenhouse gases over the long term and during the transition to the long-term allocation?

The third item on this list, background allocation, is perhaps the least familiar area of discussion; yet it provides for the basis of a fair process in determining the other three kinds of allocation. As a practical matter, background allocation does not only concern resources currently available to the negotiating parties, but also their reasonable and secure expectations of future resource availability without climate change. Two well-known models of procedural equity (Coase,

1960; Rawls, 1971) focus on the role of information – Coase positing an unrealistic situation of equal knowledge, Rawls using the veil of ignorance about winners and losers to entice participants to strike a fair bargain.

Distributional principles and allocational issues

Philosophers, political theorists, sociologists and economists have drawn up many overlapping typologies of distributional and procedural principles which could be used to elucidate alternative solutions to each of Shue's allocation problems. However, at the crudest level of aggregation, three basic sets of distributional and procedural preferences can be gleaned from the literature:

- libertarian (market utilitarian);
- contractarian (administrative utilitarian);
- egalitarian (anthropocentric and nature centric).

Three principles parallel to Shue's that can be applied to resolve the practical problems of making fair allocations of resources emerge from the work of mathematician Peyton Young (1993). These are proportionality, priority and parity. Rayner (1995) employs the same terms slightly differently from Young's usage to examine alternative proposals for allocating emissions rights.

Young uses proportionality to describe the Aristotelian principle of allocating benefits according to contribution. He distinguishes this from priority, which is allocation based on the strongest claim – for example, need in the case of kidney transplants or seniority in avoiding job layoffs. However, it seems that contribution, seniority and need are all bases for claims that are intended to be settled by administrative allocations made by an adjudicating authority. Hence, Rayner uses proportionality to indicate a distributive outcome where benefits are allocated in accordance with an administrative determination of rank, contribution or need.

This usage frees the term priority to be applied to distributional outcomes that are achieved through successful competition: in other words, first in time, first in right. This principle is well established in frontier conditions and today remains the basis of water law in the western United States. It is also a principle enshrined in the patent system, first introduced in England in 1623 (Headrick, 1990). Patents are the origin of intellectual property rights designed to concentrate the benefits of innovation on the inventor in order to stimulate and reward intellectual competition. The overall societal benefits of that increased competition are supposed to include greater volume and velocity of technological innovation and incentives for wider dissemination of innovation through benefits of scale in production.

The third principle of distribution is parity. This can be understood as the egalitarian principle of equal shares to all claimants. This is relatively simple to apply on an individualistic basis to divisible goods. However, with goods that are harder to divide, parity can lead towards the creation and maintenance of common property systems. Parity embodies the principle that each inhabitant of the earth has the right to an equal use of the atmosphere. Therefore, a fair allocation of emission rights to nation states would be one based on per capita population.

The specific proposal most widely cited for allocation in the literature is that derived directly from egalitarianism, suggesting that all human beings should be entitled to an equal share of the atmospheric resource. With minor variations, this takes two general forms: contemporary and historical per capita allocations. The net effect in all cases would be to give developing countries, with their much lower per capita emissions, a substantial excess entitlement, while the industrialized countries, with per capita emissions well above the global average, would have a deficit (Grubb, 1995).

Equal contemporary entitlement – allocation in proportion to national population – is the route proposed by Grubb (1989), Bertram (1992), Epstein and Gupta (1990), and Agarwal and Narain (1991), among others. These proposals are intrinsically an egalitarian way of deciding how to distribute a global, public and hitherto free resource. The main objections to this approach are based partly on ethical comparable-burden type arguments (since it would imply a huge adjustment burden on industrialized countries, to which they are unlikely to agree), and partly on concerns that such an approach might reward population growth. Supporters argue that any such effect is negligible compared to other factors influencing population; however, to avoid any inducement to population growth, Grubb suggests that the population measure should be restricted to population above a certain age. This modification has been criticized for discriminating against children (though, of course, like all international allocation proposals, it says nothing about discrimination within countries). Grubb et al (1992) note a wider range of possibilities for avoiding any incentive to population growth, including lagged allocation (related to population a fixed period earlier); apportionment to a fixed historical date; or the inclusion of an explicit term related inversely to population growth rate.

The international transfers implicit in equal per capita allocations would be much bigger still in proposals for equal historical or stock entitlement. Fujii (1990), Ghosh (1993) and Meyer (1995) propose that everyone should have an equal right to identical emissions regardless of country and generation. Fujii and Meyer take the fairness of this allocation as self-evident; Ghosh argues that it can be derived from a range of diverse ethical principles.

This argument has been most steadfastly articulated by the nations of the South. Yet within those nations, per capita parity is seldom, if ever, the operational distributional principle. It seems that the vast numbers close to destitution rescue the middle classes (the 'local North', as they are sometimes called) of many less-industrialized countries from the same accusations of per capita overconsumption that they themselves level at industrialized countries. Furthermore, although far from equal, the actual distribution of wealth within Northern nations is far closer to the per capita average than in Southern nations. Hence, from a Northern perspective, claims based on the parity principle that the North has a moral responsibility to take the lead in cutting consumption may not be compelling. The so-called survival emissions of poorer countries may, in practice, translate into the luxury emissions of their elites. Claims that international equity should be established on the parity principle would be seen as more compelling in the North if mechanisms were established to ensure that, for example, a carbon tax does not simply reduce the welfare of the poor in the North to benefit wealthy elites in the South (a criticism that has long been levelled at international aid). However, such mechanisms are likely to be viewed by those elites as unacceptable violations of national sovereignty.

Parity, among commentators from the less industrialized countries, is essentially an anthropocentric principle according to which the fragility of nature reflects and is exacerbated by societal inequality. However, so-called *deep ecologists* in the industrialized world go further than enjoining nature to their cause. They advocate extending the principle to non-human attributes of nature, giving equal rights to animals and legal standing to plants and morphological features of the Earth's surface (Stone, 1972). This extreme application of the parity principle has been described elsewhere as an *earthrights* model.

Parity is by no means the only basis which has been advanced in the argument over allocating emissions rights or *environmental space*. In fact, advocates of market principles, mostly in the North, have invoked the priority principle to argue that, far from incurring a debt to the South by its historic carbon emissions, the North was merely exercising its right of first in time. Such a view is reinforced by the *realist* perspective in political science and by neo-conservative economic assumptions that powerful nations will, in any case, demur at the prospect of implementing any arrangement that flies in the face of their perceived national self-interest.

According to this view, not only should past emitters be held blameless, but their current rate of emissions constitutes a status quo established by past usage and custom (Young and Wolf, 1991). Analogies are drawn with the common-law principles of adverse possession, such as squatter's rights. Ghosh (1993) not only disputes the ethical acceptability of this but also comments that pollution rights have no common law sanction. Yamin (1995), however, draws on fisheries analogies to suggest that developed countries' present use of the atmosphere could be the basis for a claim in the absence of stronger challenges to such use by developing countries.

A strict status quo allocation, proportionate to current emissions, has received widespread attention as a basis for taking a pragmatic or game theoretic approach, in that it is the default allocation that would arise in the absence of agreement, and is the only allocation basis which does not automatically impose big burdens upon the industrialized countries (Barrett, 1992).

Total cost minimization (also described as *efficiency* or a refusal to waste scarce resources) is an ethical mainstay of the priority position. Another is a prohibition on making external moral assessments of what people ought to desire. People's individual preferences are what matter, as expressed through willingness or ability to pay for what they prefer, as adjudicated by the market. Hence, this approach to climate policy emphasizes the maximization of present value utility, measured as the ratio of the future benefits of present actions.

Such analyses embody bold assumptions in the face of widespread uncertainty, some of which include emissions mitigation costs, future discounting, unmeasured externalities, a nature with no intrinsic value, and a relatively smooth period of the onset of climate change. Under this paradigm, allocation is primarily assumed to be market based, and that means allocation by preference and ability to pay.

Associated with this model is the position that delaying the response to climate change until more information becomes available is the most rational option (for example, Manne and Richels, 1992, have discussed this extensively). There will be more income in the future deriving from more productive kinds of investment, which can then be spent on more efficiently targeted abatement strategies. This position has received support, in part, because of the uncertainties associated with the scientific models. Within the same position, however, it can be argued that there is a waiting cost for delay – among other things, the future costs for greatly increased responses will be substantially higher. Shue (1994) also notes, with reference to the same position, that those who will be most able to accumulate the resources today to cope with climate change tomorrow are unlikely to be those who are the worst hit.

By analogy with the argument that patent rights result in increased general welfare through concentration of initial benefits, the priority principle is also defended on the basis that the industrial development of the North has resulted in a global increase in welfare. Therefore, a fair allocation of emission rights would be one that recognizes the historical dependence of the industrialized countries on fossil fuels and would allocate emissions rights to nation states on the basis of gross domestic product (GDP). Clearly, this approach violates the parity principle.

The third proposal is typically proportional in that it seeks to devise a formula that will transcend claims based on parity and priority. This is the *contractarian* or administrative–utilitarian approach. This allows for economic efficiency but subordinates it to a larger goal, along lines analogous to the management of a trust. The argument is that we are trustees of the Earth, and as trustees each generation has a responsibility to preserve the Earth's natural and human heritage at a level at least as good as that in which we received it (Weiss, 1989).

The trusteeship approach shares, with the advocates of extending parity principles to nature, the idea that some things are beyond price. However, its advocates are more likely than the earthrights proponents to include human artefacts in the category of priceless objects. For example, the ceiling of the Sistine Chapel, and the physical documents upon which the Magna Carta and the US Constitution are written, are examples of entrusted objects whose loss is possible but irreparable. Pricelessness, in other words, is socially constructed, with different and competing constructions at each of the three apexes of the policy space (an argument that is developed at some length in Thompson, 1979).

The model of a fiduciary trust consistent with proportionality implies (among other things): the conservation of options (defined as preserving the diversity of the natural and cultural resource base), conservation of quality (defined as leaving the planet no worse off than in which it was received), and conservation of access (defined as equitable access to the use and benefits of the legacy). It also implies a strong risk-averse strategy especially for potentially irreversible damages, and the application of precautionary and prudential principles. A notion of responsibility is also introduced (Weiss, 1989).

What is not clear in this approach is where the boundaries for this invocation of entrusting are, and this is especially important for the demarcation of what is to be kept outside of economic considerations. To put it another way: exactly what is being entrusted when we say we are committed to a planetary trust? Is this a specific trust or a general worldview within which even economics must function in a subordinate role? Does this involve some portion of the Earth's surface or certain species? Is this the capacity of the Earth to sustain itself or to sustain human life at least at our current level of flourishing?

With respect to greenhouse gas emissions reductions, the proportionality principle is essentially the contractarian approach. Such a formula might seek to combine population and GDP to provide a fair distribution of emissions rights for each nation state. For example, Wirth and Lashof (1990) proposed allocations based on an equal (50:50) mix of population and GDP, and Cline (1992) proposed an alternative allocation which consists of a weighted combination of population, gross national product (GNP) and current emissions. An approach other than historical egalitarianism in which historical contributions are considered as determining emissions allocation is a scheme examined by Grübler and Nakicenovic (1994), in which countries have to cut back from current levels in direct proportion to their responsibility for past increases (Grubb, 1995).

Grubb et al (1992), Shue (1993) and Welsch (1993) each examined a mixed system in more detail. These authors suggested an allocation which combines egalitarian and status quo/comparable burden principles in the form of a combination of population and current emissions measure, but did not specify an equal weighting of these components. Rather, they argued that the weighting accorded to population should increase over time towards a simple per capita allocation.

A wholly different quantified approach to the question of who should pay for abatement was provided by Chichilnisky and Heal (1994), who applied the concepts of classical economics to construct a strictly utilitarian formulation for maximizing global utility. They proposed that the fraction of income that each country allocates to carbon emission abatement must be proportional to that country's income level, and the constant of proportionality increases with the efficiency of that country's abatement technology. This, they argued, is the only distribution of abatement effort, for a given global total abatement and set of welfare weights, in which no country can be made better off without making another worse off – the Pareto optimality. The result, in turn, implies that the total resources each country should put towards abatement will increase as the square of the national income (Grubb, 1995).

Because it is driven by the imperative of maintaining system stability rather than by imperatives to promote growth or equality, the proportional contractarian approach is the only one that does not prefer clearly asymmetrical principles for losses and gains. Contractarians tend to favour deep pocket solutions (Calabrese, 1970) as fair mechanisms for distributing liabilities, not so much out of a desire to move towards strict equality, but to avoid destabilization of the general welfare. In this way contractarian arguments advocate redistributive mechanisms, such as taxation, to apportion liabilities in a way that seems to them to be least disruptive – not to the whole society perhaps, but certainly to those constituencies whose stability they see as critical.

Libertarians (market utilitarians) tend to favour loss-spreading in which market systems, such as insurance and reinsurance, determine who bears losses (Calabrese, 1970). When pure market solutions are not available, advocates of this position attempt to reproduce what the market would have done if it had not been impeded by high information and/or transaction costs. For example, the Price-Anderson Act was designed to limit the liability carried by nuclear plant operators in the United States in favour of diverting the costs to the population in general. This is asymmetrical to the principle that allows the individual risk initiator to collect any gains that may accrue.

In contrast with both of the broadly utilitarian approaches, egalitarians seek a moral determination of liability that appeals directly to egalitarian values. This is a strict-fault system (Calabrese, 1970) that makes those who are seen as responsible for imposing the risk directly responsible for the costs of making the polluter pay.

The Organisation for Economic Co-operation and Development (OECD) formally adopted a form of the polluter pays principle in 1974 as a guide to environmental policy in stating that, if measures are adopted to reduce pollution, the costs should be borne by the polluters. This is an economic principle which says that polluters should bear the cost of abatement without subsidy. The principle does not explicitly state that all polluters of a common pollutant should pay in proportion to their emissions, and in fact the literature seems remarkably opaque about how the principle should be applied in a context such as climate change. The principle clearly points towards responsibility-based rather than burden-based criteria, proportional in some way to emissions. The polluter should pay: but on what basis, who should receive payment and for what purpose?

A critical distinction that is rarely clarified is whether the principle applies to gross payment or net payment. Burtraw and Toman (1992) assume that it applies to net payment, so that each country should pay for abatement in proportion to its contemporary emissions; this, they note, would be regressive against national income, a characteristic that is bound to spark developing country opposition. Chichilnisky and Heal (1994), among others, have pointed out that such allocation of payment may be neither efficient nor fair. Other authors assume that the principle means that gross payments should be proportional to emissions, leaving open the matter of how the resulting revenues should be distributed as a separate question of efficiency and equity; and they also consider different bases for payment (Grubb, 1995).

By holding the polluter strictly liable, egalitarians aim to eliminate incentives to cut corners on safety, or even to continue with the activity at all. The strong preferences of egalitarians for allocating costs to guilty parties is asymmetrical to their communal principle of the broadest spread of gains. Therefore, the argument in favour of parity in the distribution of emissions rights goes hand in hand with the argument that the North should bear the costs of climate change policies because it is responsible for the bulk of historical greenhouse gas emissions. These preferences for distributional outcomes are summarized in Table 2.1.

Table 2.1 Outcome fairness and asymmetry of losses and gains according to three ethical positions

	Libertarian	Contractarian	Egalitarian
Principle for gains	Priority	Proportionality	Parity
Principle for losses	Loss spreading	Deep pocket	Strict fault
Outcome for gains	Narrowest	Greatest good	Broadest
Outcome for losses	Broadest	Least harm	Narrowest

Procedural fairness in climate change

The issue of procedural fairness – whether the means by which an outcome is reached is considered to be fair and reasonable – has been studied by interpretive social scientists analysing societal responses to technological hazards. This approach focuses on preferred procedures for obtaining consent to risk (MacLean, 1980; Rayner and Cantor; 1987). Institutional preferences for valid principles of consent are summarized in Table 2.2.

Table 2.2 Valid consent as framed by three ethical positions

	Libertarian	Contractarian	Egalitarian
Consent	Revealed	Hypothetical	Explicit

With respect to consent, decision-makers in a market individualist context favour a *revealed preference* approach (Thaler and Rosen, 1975) – sometimes called implicit consent (MacLean, 1980). For example, if the price differential between three ladders reflects only the degree of safety built into each one and a consumer selects the mid-priced ladder, that consumer is deemed to consent to the additional risk that results from not purchasing the costliest ladder. At the same time, the consumer reveals that he or she does not consent to the increased risk that would result from purchasing the cheapest ladder. This principle allows market forces to determine planning priorities and the degree of risk which people are prepared to accept. The rationale is that people's preferences for one solution or another will be reflected accurately in how they spend their money (which, of course, assumes they have some).

The contractarian principle for obtaining consent is sometimes called *hypothetical consent* (Rawls, 1971). The citizen is assumed to have entered into a contract with the decision-making institution, where he or she may be deemed to assent to decisions made through rational procedures of that institution, even though he or she may not like the particular outcome. For example, we pay income tax because we accept the legitimacy of the government's claim to levy taxes rather than because we agree with the amount that it charges us or the particular pattern of spending it chooses. Contractarian appeals to procedural fairness in obtaining consent are likely to be compared to the constitutional procedures for decision-making.

Explicit consent is the only legitimate form of consent for egalitarians. The use of any surrogate for consent undermines the basic premise of egalitarianism, that all are the same and have equal say. This gives rise to particular difficulties in assigning responsibility to the present generation for the acts of our forebears as well as for obtaining consent from future generations.

Consider the claim that the North bears a special obligation to pay for climate policies based on the historical dimension of global resource use. There are two plausible lines of reasoning here. One is that children do indeed inherit the liabilities of their parents along with their assets. The other is that the historical behaviour of the North has created a current condition of structural dependency of the South upon the North.

Historical responsibility as an equity principle has strong support in the literature and politically in developing countries, but there are also valid counter-arguments. These include:

- ignorance of past generations about the consequences of their actions;
- ambiguities in tracing responsibility for emissions; and
- the fact that the benefits of emissions have spread beyond the emitters.

Smith et al and Ghosh reject such objections as being partly inaccurate, but mostly irrelevant to the Fujii (1990) principle of equal historical per capita entitlement, which is based not upon fault, blame or compensation but upon an egalitarian principle of access. In fact, the disagreements illustrate several different dimensions to the debate about historical responsibility.

With respect to the principle of historical obligation, it is easy to reconcile the principle that the North has liabilities towards the South, based on past resource extraction, if the relevant entities of the North and the South are judged to be legally immortal hierarchical institutions, such as corporations or nation states. This would be quite consistent with the *hypothetical consent* principle, where the individual is deemed to consent to the decisions of legitimate institutions, even though he or she might individually dissent. However, the preferred asymmetry of losses and gains enables the egalitarian to reconcile the contractarian argument about inherited liabilities with the egalitarian principle of parity advocated in the per capita allocation of greenhouse gas emissions. This implies that the claims of individuals take precedent over those of other entities such as states. Under egalitarian principles, an individual can only incur a debt by *explicit informed consent* and cannot be held liable for the debts of his or her forebears.

Smith et al (1993) propose that responsibility for paying should be determined on the basis of the *natural debt* – in proportion to total cumulative emissions since a specified date. Because this in itself would result in all countries bearing some responsibility for paying (though very much less for developing countries), they modify this by suggesting a lower threshold for *basic needs* emissions, consistent also with the arguments of Agarwal and Narain (1991) and others.

But how far back should emission estimates go? Should the natural decay (reabsorption) of emissions be taken into account, and if so how? Which gases should be included? How should the emissions be related to scale (for example, cumulative population or current population)? In part, these complexities reflect different potential definitions of the natural debt concept. When assessing responsibility indices, for illustration, Smith et al consider total industrial carbon dioxide (CO_2) emissions since 1950 by country, with a range of lower threshold levels. However, at the same time as espousing the Fujii formulation of intergenerational egalitarianism, Ghosh (1993) criticizes the natural debt concept as a basis of historical responsibility on the grounds that it is an abstract, environment-centred focus that does not acknowledge responsibility to others for one's actions and does not relate to fairness among human beings (Grubb, 1995).

EQUITY, INSTITUTIONS AND SOLIDARITY

What is fair may be the subject of disagreement, but the demand for fairness only arises because, as John Donne put it, 'No man is an island'. It is very hard to imagine what fairness would mean if we did not live and work together in families, communities, firms, nations and other social arrangements which persist over time. The whole issue of fairness arises out of the establishment of public, that is, shared, expectations for the conduct of community relations (procedural equity) and the distribution of rights over resources within and among communities (distributional equity). In other words, fairness is integral to establishing and maintaining social solidarity at every level of social institutions, from the micro to the macro, from the local to the global (Rayner, 1995). Protest and defection from institutions result when public expectations for procedures, or for the outcomes of allocations, are repeatedly violated or people cannot be persuaded to embrace emergent alternatives (Hirschman, 1970).

Issues of equity in climate change serve to highlight the central importance of the concept of social solidarity to understanding the social and political discourses about climate. They suggest that these discourses are inextricably institutional in nature and are not simply the rational expressions of preformed individual preferences as assumed by the self-interest paradigm that has dominated economic and political analysis throughout most of the 20th century.

Since World War II, political theorizing has been dominated by so-called non-institutional perspectives. Non-institutional theories can be characterized as:

- methodological individualist – political outcomes are more likely to be seen as resulting from the sum of individual actions, rather than being attributed to rules or structural factors;
- utility maximizing – individuals are more likely to be seen as acting in accordance with what they determine to be their best interest, rather than as responding to perceived responsibilities and roles;
- instrumentalist – politics is more likely to be seen as action (the making of decisions and the distribution of goods and services) rather than as a process of creating meaning;
- functionalist – the historical development of a society is likely to be seen as a process of successful adaptation such that, over time, it reaches its own unique and stable configuration; dysfunction, instability and similitude are not the focus of attention (March and Olsen, 1989).

Such theories include those concerned with competitive, rational agents (realism, rational choice, non-repetitive game theory and its variants, for example) and those concerned with the temporal sorting of problems, solutions, decision-makers and choices, such as Kingdon's (1984) *stream* model of the flow of policy events.

However, by these very assumptions, non-institutional approaches identify themselves with the individualist libertarian ethical stance which underpins market-style institutional arrangements. They fit comfortably with the benign view of nature (Thompson, 1987). As long as we all do our own exuberant individualistic things, a hidden hand will lead us to the best outcome. From this point of view, we interfere with this process at our peril. Denying the institutional nature of human interaction is itself an institutional stance.

Managing climate change has to take place through institutional arrangements. Any attempt to conceptualize climate change will involve an institutional framework of one kind or another. Similarly, how responses to climate change are identified, evaluated and enacted can only take place through the medium of institutional behaviour.

Three forms of solidarity

More than 150 years of social theory – spanning legal history, sociology, anthropology, psychology, economics and political science – indicate that institutions (solidarity) can be built in many ways and across different scales, from families to federations of nation states. This variation has often been presented as an evolutionary dichotomy – an inexorable historical journey from state A to state B – but (as we will see) with the theorists often selecting their As and their Bs very differently.

Dichotomous distinctions have proven to be very durable in the history of social theory. In the mid 19th century, the legal historian Sir Henry Maine (1861) distinguished social solidarity based on *status*, in which actors know their place in hierarchical structures which stem from the idiom of the family, from solidarity based on *contract* in which agents freely associate by negotiated agreement. Later in the century, the German sociologist Ferdinand Tönnies (1887) distinguished between *Gemeinschaft*, where societies are bound by ties of kinship, friendship and local tradition, and *Gesellschaft*, where social bonds are created by individualistic competition and contract. At the turn of the century the French anthropologist Emile Durkheim (1893) distinguished human societies based on *mechanical solidarity*, in which agents bind themselves to others on the basis of sameness, from those built upon *organic solidarity*, in which agents are bound together by the interdependence of specialized social roles. Each of these grand dichotomies was viewed by its author in evolutionary terms which continue to resonate in contemporary social theorizing, such as the ideas of Bennett and Dahlberg (1990) who, echoing Durkheim, detect in the development from preindustrial to industrial society a shift from *multifunctionalism*, where everyone can do everything, towards *specialization*.

For the most part, however, more recent approaches dispense with the unidirectional evolutionary assumption. Educational sociologists have identified *positional* families, in which behaviour is regulated by appeals to hierarchical authority, and *personal* families, in which behaviour is regulated by appeals based on individual preferences (Bernstein, 1971). Major contemporary political scientists and economists such as Charles Lindblom (1977) and Oliver Williamson (1975) focus on the different characteristics and dynamics of coexisting and competing social systems based on the social bonds created through participation in *markets* and those based on the solidarity of *hierarchy*.

Parsimony requires that we reduce this rich diversity to the minimum number of basic modes or patterns of solidarity that can be distinguished

Table 2.3 Characteristics of three kinds of social solidarity as described in classic social science literature

Market	Hierarchical	Egalitarian
Gesellschaft	Gemeinschaft	Gemeinschaft
Organic solidarity	Organic solidarity	Mechanical solidarity
Specialized roles	Specialized roles	Multifunctional roles
Personal authority	Positional authority	Personal authority
Contract relations	Status relations	Status relations

usefully. It is also desirable to have a basis for distinction that would hold across the board, from micro to macro scales of social organization. The grand dichotomies overlap a good deal, but they are far from perfectly congruent and seem not to be reducible to fewer than three (see Table 2.3).

Firstly, solidarity can be expressed through the *market*, characterized by the features of individualism and competition associated with *Gesellschaft*. Solidarity is achieved in two ways, most obviously through *contracts*, but also through individual consumption choices that establish identity with fellow consumers and differentiation from those who follow different consumption patterns. As manifestations of *Gesellschaft*, market forms of solidarity are directly orthogonal to both of the other basic modes of solidarity, described below, which share the stronger community boundaries typical of *Gemeinschaft* solidarity.

Secondly, solidarity can be expressed through orderly differentiation in *hierarchies*, the rules for which establish identity through careful gradations of *status* based on explicit characteristics such as age, gender, educational attainment and professorial rank. This form of *positional* authority is directly orthogonal to the emphasis on *personal* freedom shared by both markets and the third form of social solidarity.

Thirdly, solidarity can be expressed through *egalitarian* homogeneity – that is, by operating rules of equality that keep each participant at the same *status*. In this respect, egalitarianism is a manifestation of *mechanical* solidarity and *multifunctional* roles that are directly orthogonal to both hierarchies and markets, which both favour *organic specialization* of labour.

This synthesis creates a two-dimensional space within which multiple possibilities for institutions can be located. This has several methodological and pragmatic advantages over dichotomous frameworks. It systematically encompasses the spectrum of dichotomous distinctions that, over a century and a half, have informed a wide range of empirical social sciences. This inclusiveness suggests that the three basic types are fairly robust.

The endogenous dynamic qualities of the two-dimensional framework are also methodologically appealing. Each kind of solidarity only exists in opposition to the other two, which means that instability and conflict are inherent to the framework, as they are in real life, and do not require the action of an exogenous agent for changes in social organization or the values that support it. The potential for endogenizing social change may have important implications for long-term policy modelling, which presently is unable to deal well with changes in societal preference functions such as the demographic transition.

We have thus far explored ethical principles as building different kinds of solidarity among the living. Because of the very long timescales of climate change issues, we now turn to the issue of building solidarity between the living and the dead and the unborn: the issue of intergenerational equity.

Solidarity and intergenerational equity

The argument about historical debt invites us to address intergenerational equity as an issue of solidarity across generations – how we bind ourselves to our ancestors and to our descendants. Integrated assessment models do not explicitly deal with this issue, except to choose one discount rate. Table 2.4 summarizes the effects of each kind of social solidarity on expectations of the future, responsibility to future generations and determination of the discount rate or rates.

Table 2.4 Time, intergenerational responsibility and discount rates as framed by three kinds of solidarity

	Market	Hierarchical	Egalitarian
Time perception	Short term	Long term	Compressed
Intergenerational responsibility	Weak	Balanced	Strong
Discount rate	Diverse high	Technically calculated	Zero or negative

In market institutions, expectations of the future are likely to be strongly focused on deadlines (Rayner, 1982). Competitive success depends largely on timing: planning for shifting market tastes, clinching deals at the right price, meeting delivery deadlines, or knowing when to sell. The emphasis is on short-term expectations and immediate returns on activities and investments. Hence, market institutions pay little heed to intertemporal responsibility. They tend to assume that future generations will be adaptive and innovative in dealing with the legacy of today's technology, just as our generation has had to be in response to the legacy of the industrial era. So far as consent is concerned, it is assumed that future generations will make decisions on current market conditions and will therefore accept similar conditions of their predecessors. The emergence of future liabilities can be left to market forces when they occur and will, in fact, provide the stimulus for future enterprise. Under these conditions different discount rates apply simultaneously for different goods or at different times for the same good. The discount rates also tend to be high.

In hierarchical institutions, history is strongly differentiated. Anniversaries of great events in the past are celebrated collectively and provide models for discriminating epochs of the future. Clear recognition of age cohorts and generations, which are the basis for establishing seniority in the present, also engender clear expectations of an ordered future. The regimes of distinguished leaders (whether kings or company directors) also contribute to an ordered expectation of the future. Intergenerational responsibility therefore tends to be strong but balanced by the needs of the present. It is also likely to be safeguarded by the longevity of institutions. Consent is based on the assumption that future generations will recognize the legitimacy of present institutions. The apparent discount rate, therefore, tends to be lower than in the instances where market solidarity applies. Furthermore, hierarchies are the most likely of the three kinds of solidarity to be concerned with the bureaucratic determination of a standardized rate that can be applied across the board.

In egalitarian groups also, history tends to be viewed as epochal; however, because of the problems of resolving disputes in institutions that are reluctant to recognize dispute resolution by claims to seniority, competitive leadership or established procedures, such groups are prone to frequent schism (Rayner, 1986). As a result, the group's crusading mission may lead to a sense of historical self-importance that results in the view that the present epoch is a decisive historical moment. Under these conditions, intergenerational responsibility is very strong, but trust in formal institutions is weak. If consent cannot be obtained from future generations, and our descendants cannot force long-dead decision-makers to pay for their errors, then we have no right to accept risks on behalf of those descendants. Under these conditions the apparent discount rate used for environmental and intergenerational calculations is very close to zero, possibly even negative.

The different perceptions of time and expectations of the future engendered by each kind of solidarity seem to be critical factors in their perceptions of the costs and benefits of climate change policies. The high discount rates and low levels of intertemporal responsibility typical of market solidarities combine to focus the attention of decision-makers on current costs. Furthermore, the desired asymmetry of losses and gains leads those same decision-makers to defer costs into the future where they may be capable of being dispersed more widely throughout society. Hence, there is a strong incentive in market solidarity to postpone policy responses to the threat of climate change.

The low discount rates of egalitarians, combined with high levels of intertemporal responsibility and the impossibility of obtaining explicit consent from future generations, focus decision-makers' concern on potentially high, possibly catastrophic, future costs. This provides a powerful incentive to take action now which, in keeping with the desired asymmetry of costs and benefits, also places the burden on the parties actually responsible for the current emissions that may affect future generations. Hence, egalitarian solidarity leads to emphasis on rapidly implementing policies designed to prevent the onset of climate change.

The technocratic approach to discounting and the overriding concern for system maintenance in hierarchical solidarity lead towards a middle course – limiting responses only to the 'no regrets' strategy of implementing climate change policies that make sense for other reasons while attempting to improve understanding of the extent of present and future costs, and when they might arise.

A Two-Dimensional Map of Institutional Discourse and Human Values

In the preceding sections we have summarized a wide range of social science insights into the three most basic ways in which people bind themselves to each other in social institutions and, in so doing:

- establish principles for intergenerational equity;
- diagnose the underlying social causes of climate change;
- develop preferences for distributional equity; and
- establish principles of procedural fairness.

These summaries provide us with a basic two-dimensional map of the climate change discourse as framed by the diversity of human values (see Figure 2.4). The triangle is equivalent to the territorial borders of a conventional geographical map. Using the same outline, separate maps can be drawn for each element of equity, just as atlases contain maps of precipitation, vegetation, population and other variables of interest that are superimposed upon the land surface. Often several maps can be overlaid to give a composite picture as in Figure 2.4. This analogy with geographical maps is helpful precisely because such maps always tie the variable of interest to a particular location having specific geological characteristics that support the variable. In other words, precipitation, vegetation, population, etc are not free floating, but tied to specific places on the land surface. Similarly, the map of human values is not just a free-floating suite of policy biases and ethical options. It locates packages of values and ethical priorities in specific places in the social landscape – that is, within the discourse among institutions exhibiting various mixtures of market, hierarchical or egalitarian solidarity.

The analogy with geographical maps is also useful in that maps can be drawn at different scales. Similarly, the map of discourse and human values can be drawn at a variety of scales. The family, the workplace and the community all depend upon the creation of solidarity among individuals. The corporation, the nation state or the global community of nations are examples where solidarity is built among larger aggregations of people bound to each other in various ways. Ultimately, the issue of climate change, in particular, and global sustainable development, in general, is a challenge to create and maintain solidarity at all of these levels. The two-dimensional map provides us with a richer analytical framework than do one-dimensional approaches for understanding and intervening in ethical disputes when forming treaties, implementing national policies, shaping consumer demand and modifying behaviour at the microlevel of firms and families.

Another feature of geographical maps drawn to scale is that they enable a navigator to locate his or her position in relation to other positions and to measure the distance from various objectives. Systems of mathematical measurement of market, hierarchical and egalitarian solidarities already exist at the level of communities and organizations (Gross and Rayner, 1985) and of households (Dake and Thompson, 1993). Similar measurement schemes have been adapted for measuring solidarities among larger units such as nation states (Grendstad and Selle, 1997). Negotiators can use the map of values to

Hierarchical institutions
Myth of nature: perverse/tolerant
Diagnosis of cause: population
Policy bias: contractarian
Distribution: proportionality
Consent: hypothetical
Liability: deep pocket
Intergeneration responsibility: present = future
Discounting: technical standard

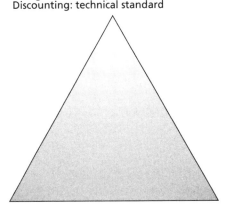

Market institutions
Myth of nature: benign
Diagnosis of cause: pricing
Policy bias: libertarian
Distribution: priority
Consent: revealed
Liability: loss-spreading
Intergeneration responsibility:
    present > future
Discounting: diverse/high

Egalitarian institutions
Myth of nature: ephemeral
Diagnosis of cause: profligacy
Policy bias: egalitarian
Distribution: parity
Consent: revealed
Liability: strict fault
Intergeneration responsibility:
    future > present
Discounting: zero/negative

Source: Rayner, 1995

Figure 2.4 A two-dimensional map of institutional discourse
and human values

locate their own positions, to select the direction in which they want to move in any dimension of equity, and to measure, on multiple criteria, the distance between their own positions and the goals of other participants in a negotiation process.

At the very least, use of the map could lead to technical improvements in the kind of benefit-cost analyses already being performed within the IPCC framework. For example, the map of human values rests on the processes of creating and maintaining social solidarity in both space and time. As a result, it provides a theoretical basis for developing predictive models of institutional differences in principles of intergenerational obligation and social discount rates, which are key to the success of welfare economics, but into which welfare economics provides no insight.

In fact, welfare economics as a prescriptive value system has its own place on the two-dimensional map. The dominant concern for efficiency, embodied in the drive to achieve Pareto optimality, is characteristic of market solidarity. However, within welfare economics, the market orientation is modified by hierarchical concerns to apply the efficiency criterion to administrative determinations of fairness. This locates welfare economics along the left side of the triangle, between market and hierarchical approaches. Therefore, the triangular map graphically reminds us to consider the alternative utility functions: those that value equality over efficiency and refuse to discount future costs. While some economists may be tempted to argue that such egalitarians are too few to matter in the grand scheme of things, others who study the dynamics of scientific and technological disputes will attest to the effectiveness of such parties in slowing or preventing the implementation of a variety of projects and programmes, including hazardous waste facilities, nuclear power programmes, genetic engineering research development and marketing (Johnson and Covello, 1987) and the Brent Spar's burial at sea.

By focusing on the construction and maintenance of institutional solidarities, rather than on the preferences of individuals, the map also provides a basis for developing models of changes in societal preference functions. Furthermore, by making the form of solidarity the unit of analysis, and by allowing an individual to be part of several solidarities, we can appreciate how it is that individuals can make radical shifts in their individual preference functions. We no longer have to resort to the conclusion that they are either irrational or acting in bad faith.

For example, a negotiator from a developing country elite may invoke the parity principle in international negotiations, the proportionality principle in apportioning budgets in his or her domestic bureaucracy, and the priority principle in making his or her own domestic private-sector investments. Rather than cynically switching values in the pursuit of naked self-interest, the map suggests that the same individual may simply be involved in building different kinds of solidarity in different spheres of activity.

Each package of principles for defining needs, preferences for distributional outcomes and procedures, responsibilities to future generations, and obligations with respect to nature is stabilized institutionally when creating social solidarity. By focusing on the form of solidarity, and not on the psychophysiological entity, we are able to cope with that most micro of institutional realities: the *dividual*, a term coined by Marriott (1967) to convey the point that there is nothing socially indivisible about the individual.

Furthermore, as Rose (1990:927) observes, while there is good consensus among welfare economists on Pareto optimality as the best efficiency criterion, 'there is no consensus on a "best" equity principle'. The two-dimensional map enables us to refine this observation; that is, we probably should not even attempt to find a best-equity principle, but should focus instead on achieving practical agreement about joint action among parties upholding quite different, even incommensurable, principles of equity. And, of course, embedded in the Pareto principle is one fairness principle: priority. In other words, integrated assessment should shift our sights from a *technocratic* goal of providing decision-makers with the best possible prescription for fairness towards the more modest goal of providing decision-makers with the best tools for essentially *political* negotiation among competing prescriptions. As Rose (1990:934) concludes:

> *The fact that there are several alternative definitions of equity should not give cause for dismay but, rather, should stimulate further study. In principle, nearly all nations subscribe to some*

concept of fairness, and thus its potential as a unifying principle should not be neglected.

The view that equity can be a rallying point in climate negotiations is echoed by several authors, including Hahn and Richards (1989), Toman and Burtraw (1991), and Rose and Stevens (1993). A concept of fairness that acknowledges diversity across more dimensions than distributional outcome alone seems more capable of engaging diverse actors than a one-dimensional concept that may require participants to violate their values and preferences on other dimensions of equity.

This conclusion is consistent with the advice of decision theorists that parties seeking agreements on actions should include a wide range of dimensions in their negotiation (Fisher and Ury, 1981; Raiffa, 1982). To design *win–win* solutions in negotiations, we need to be able to recognize what counts as a win to the parties involved. Winning consists of more than just the bottom line from a single transaction; building communities at the local and global scales is a highly valued goal of human behaviour. For instance, the US State Department frequently argues that participating in international environmental agreements enhances US leadership and credibility in the international community. This desire to promote goodwill and solidarity among nations may well override concerns for the domestic costs of participation (Hahn and Richards, 1989). The map of human values provides a systematic tool for exploring multiple dimensions of fairness in the search for practical agreements.

DIRECTIONS FOR INTEGRATED ASSESSMENT

We have described a moral landscape and an approach to mapping it that presents a picture of the equity concerns that are likely to influence the conduct of climate negotiations and efforts to implement climate change policies. This landscape is more complex than the one-dimensional analytical framework that currently dominates discussions of equity. However, it remains reasonably parsimonious and has clear practical implications for integrated assessment of climate change and climate change policies.

In navigating this moral landscape, we recognize that climate change provides an arena for debating a wide variety of social, economic and political issues that society in general, and integrated assessment in particular, find difficult to address directly. These include the unequal distribution of wealth within and among nations and the tension between the imperatives of independence and interdependence at all levels of social organization. Much of the debate about equity in climate change mitigation is an extension of the broader debate about international economic development and political empowerment.

Clearly, there is a social benefit to be obtained from the existence of an arena in which potential changes in the socioeconomic and political status quo can be explored as deriving from natural imperatives rather than from human agency. This enables parties to advance agendas for change without directly and immediately threatening deeply entrenched political and financial interests.

But the situation also presents potential dangers for human society. On the one hand, it is plausible that the opportunity costs of debating significant social and economic change in a surrogate arena may reduce our capacity to make desirable changes. For example, if we allocate significant economic and political resources to mitigating climate change as a way of enhancing the development of less-industrialized countries, we may be reducing the level of resources actually available to fight poverty, hunger and ignorance. On the other hand, our use of the opportunity that a potential natural crisis provides for social reform may lead us to ignore or override signals from the natural system that nature is, indeed, about to use its veto over human behaviour.

These are important questions that are seldom addressed directly by policy-makers engaged in the climate change discourses. Unfortunately, policy- and decision-makers often seem to fear that making these implicit dimensions of the discourse explicit will undermine political support for their own positions. If so, there is no basis for such fear in the kind of social science analysis presented here. Recognizing epistemological and ethical diversity does not lead inexorably to agnosticism and political paralysis. Rather, it provides a realistic perspective from which to participate in debate with a heightened ability to listen to, and understand, the arguments and standpoints of other participants as well as to be aware of the role played by one's own institutional bias.

Furthermore, climate change is widely seen as part of a broader problem of humankind's relationship with nature. However, views of what that relationship should be are diverse and deeply rooted in our social relationships – in social solidarity. How we bind ourselves to each other shapes the way we bind ourselves to nature. Social and cultural variables of network density and interconnectedness, as well as rule sharing, may better account for variations in environmental perceptions and behaviour than standard demographic variables such as age and sex.

Important policy differences with respect to diagnosis and treatment of climate change reflect these differences in ways that fundamentally shape political and administrative solutions. The conceptual map of human values is a useful and parsimonious device, both for social scientists to identify and track the strength of support for alternative positions and for policy-makers identifying opportunities for effective intervention in the debates.

The insights from the conceptual map of human values can be incorporated into integrated assessments in a variety of ways. These could be making incremental technical adjustments or providing inputs to existing models, or they could involve fairly radical rethinking of the goals and methods of integrated assessment as a societal decision aid.

At the most modest level, the map could be used to adjust inputs to existing deterministic models such as the GCAM (Edmonds et al, 1994). Although this has not been attempted, the model could be run with different discount rates reflecting the equity assumptions and preferences of hierarchical, market and egalitarian institutions. More radically, the three approaches to equity (defined broadly as what is right and what is natural) could be used to reconstruct existing models to incorporate variety, as has been done by van Asselt et al (1995) with the TARGETS model. Even more radically, the map can be used to develop non-deterministic models in which the three institutional worldviews compete – an approach already taken up by Marco Janssen (1996). The most dramatic opportunities for this approach might be in integrated assessment activities designed to reembed computer models in a social context for decision-making that is more broadly defined than the current mode of science speaking truth to power (Pahl-Wöstl, 1998), such as the ULYSSES Project led by Carlo Jaeger.

The map of human values highlights the expert/lay dichotomy at the heart of integrated assessment modelling. As currently practised, integrated assessment models assume that one homogeneous group of experts generates and transfers knowledge to another homogeneous group of decision-makers. This dichotomy obscures significant variation in the perceptions and preferences of both. It also obscures the fact that real people are not consistently experts or lay people. There are no universal experts and, in the civic arena, even the most modest lay person has some relevant expertise. Relevant knowledge brought to bear in the climate discourses is not comprised solely of scientific facts about climate chemistry, dynamics and impacts, but also derives from various experiences of social change and societal responses to natural change.

The lay/expert dichotomy also structures communication as a unidirectional process in which expert knowledge is passed to the public, either to alleviate its ignorance or to redress its misperceptions. In this mode decision-makers are frequently stopped in their tracks by recalcitrant populations who rightly insist that they have not been heard and that their expertise (what anthropologists call local knowledge) has been ignored. The suggestion that expert discourses are structured by the same elements of social organization as are lay discourses redirects efforts at communication from simply overcoming ignorance to creating shared frames of reference and opportunities for shared action. Public information campaigns which assume that the discrepancies between lay and expert accounts of climate change are simply due to knowledge deficiencies are bound to fail. Effective communication about climate change issues requires understanding of the frames of reference being used by all participants.

Climate discourses are complex and turbulent. Many voices join in and they are frequently inconsistent, even self-contradictory, but not randomly so, nor in a way that can simply be ascribed to naked self-interest. However, plurality is not always easy to discern because the vocabulary of one voice or another may define the debate, for a time, on its own terms. Under these conditions of *hegemonic discourse* (Rayner, 1992), dissenting voices may not be clearly heard because they are forced to use the terminology of the hegemonic voice. For example, it is very difficult to reject the egalitarian claims of equal per capita emissions rights in international negotiations, even though the resulting allocation of those rights to nation states would have no discernible effect on promoting individual equality. The point here is not that the egalitarian position is necessarily wrong, but that society's ability to explore alternative formulations and solutions is constrained under conditions of hegemony and, as a result, important opportunities for solutions may be lost.

REFERENCES

Agarwal, A and Narain, S, 1991, *Global warming in an unequal world*, Centre for Science and Environment, New Delhi

Barrett, S, 1992, 'Acceptable' allocations of tradeable carbon emission entitlements in a global warming treaty, in: *Combating global warming*, UNCTAD, Geneva

Bennett, J W and Dahlberg, K A, 1990, Institutions, social organization, and cultural values, in: Turner et al, *The Earth as transformed by human action*, Cambridge University Press, Cambridge

Bernstein, B, 1971, *Class, codes, and control*, Vol 1, Routledge and Kegan Paul, London

Bertram, G, 1992, Tradeable emission permits and the control of greenhouse gases, *Journal of Development Studies* **28**(3):423–446

Bloomfield, B P, 1986, *Modelling the world: the social constructions of systems analysts*, Blackwell, Oxford

Bullard, R D, 1994, Unequal environmental protection: incorporating environmental justice in decision making, in: A Finkel and D Golding, eds, *Worst things first: the debate over risk-based national environmental priorities*, Resources for the Future, Washington, DC

Burtraw, D and Toman, M A, 1992, Equity and international agreements for CO_2 constraint, *Journal of Energy Engineering* **118**(2):122–135

Calabrese, G, 1970, *The cost of accidents*, Yale University Press, New Haven

Chichilnisky, G and Heal, G, 1994, Who should abate carbon emissions? An international viewpoint, *Economic Letters* **44**:443–449

Coase, R H, 1960, The problem of social cost, *Journal of Law and Economics* **3**:1–44

Cline, W, 1992, *The economics of global warming*, Institute for International Economics, Washington, DC

Cotgrove, S, 1982, *Catastrophe or cornucopia: the environment, politics and the future*, John Wiley, New York

Dake, K and Thompson, M, 1993, The meanings of sustainable development: household strategies for managing needs and resources, in: S D Wright, T Dietz, R Borden, G Young, G Guagnano, eds, *Human ecology: crossing boundaries*, Fort Collins, CO

Durkheim, E, 1893, *De la division du travail sociale: étude sur l'organization de societés superieurs*, Alcan, Paris

Edmonds, J A, Wise, M and MacCracken, C N, 1994, *Advanced energy technologies and climate change: an analysis using the Global Change Assessment Model (GCAM)*, PNL 9798, Pacific Northwest National Laboratory, Richland, WA

Elster, J, 1992, *Local justice: How institutions allocate scarce goods and necessary burdens*, Russell Sage Foundation, New York

Epstein, J and Gupta, R, 1990, *Controlling the greenhouse effect: five global regimes compared*, Brookings Institution, Washington, DC

Fisher, R, and Ury, W, 1981, *Getting to yes: negotiating agreement without giving in*, Houghton Mifflin, Boston

Fujii, Y, 1990, An assessment of the responsibility for the increase in the CO_2 concentrations and intergenerational carbon accounts, Working paper WP–05–55, IIASA, Laxenburg, Austria

Fulkerson, W, Cushman, R M, Marland, G, Rayner, S, 1989, International impacts of global climate change, Testimony to House Appropriations Subcommittee on Foreign Operations Export Financing and Related Programs, ORNL/TM–11184, Oak Ridge National Laboratory, Oak Ridge, TN

Ghosh, P, 1993, Structuring the equity issue in climate change, in: A N Achanta, *The climate change agenda: an Indian perspective*, Tata Energy Research Institute, New Delhi

Grendstad, G and Selle, P, 1997, Cultural theory, postmaterialism and environmental attitudes, in: R Ellis and M Thompson, eds, *Culture Matters*, Westview, Boulder, CO

Gross, J and Rayner, S, 1985, *Measuring Culture*, Columbia University Press, New York

Grubb, M, 1989, *The greenhouse effect: negotiating targets*, Royal Institute of International Affairs, London

Grubb, M, 1995, Seeking fair weather: ethics and the international debate on climate change, *International Affairs* **71**(3):463–496

Grubb, M, Sebenius, J, Magalhaes, A and Subak, S, 1992, Sharing the Burden, in: I M Mintzer, ed, *Confronting climate change: risks, implications, and responses*, Cambridge University Press, Cambridge

Grübler, A and Nakicenovic, N, 1994, *International burden sharing in greenhouse gas reduction*, RR-94-9, IIASA, Laxenburg, Austria

Gudeman, S, 1988, *Economics as culture: models and metaphors of livelihood*, Routledge and Kegan Paul, London

Hahn, R W, and Richards, K R, 1989, The internationalization of environmental regulation, *Harvard International Law Journal* 30(2):421-446

Headrick, D R, 1990, Technological change, in: Turner et al (1990)

Heineman, R A, Bluhm, W T, Peterson, S A and Kearny, E N, 1990, *The World of the Policy Analyst: Rationality, Values, and Politics*, Chatham House Publishers, Chatham, NJ

Hirschman, A O, 1970, *Exit voice, and loyalty: responses to declines in firms, organizations and states*, Harvard University Press, Cambridge, MA

Holling, C S, 1986, The resilience of terrestrial ecosystems, in: W C Clark and R Munn, eds, *Sustainable Development of the Biosphere*, Cambridge University Press, Cambridge

Ingold, T, 1993, Globes and spheres: the topology of environmentalism, in: K Milton, ed, *Environmentalism: The view from anthropology*, Routledge, London

Janssen, M, 1996, Meeting targets: tools to support integrated assessment modelling of global change, PhD thesis, University of Maastricht

Johnson, B B and Covello, V T, eds, 1987, *The social and cultural construction of risk: essays on risk selection and perception*, Reidel, Dordrecht

Kasperson, R and Dow, K, 1991, Development and geographical equity in global environmental change: a framework for analysis, *Evaluation Review* 15(1):149-171

Katama, A, ed, 1995, *Equity and social considerations related to climate change*, Papers presented to IPCC Working Group III Workshop, ICIPE Science Press, Nairobi

Kempton, W, Boster, J S and Hartley, J A, 1995, *Environmental values in American culture*, MIT Press, Boston

Kingdon, J, 1984, *Agendas, alternatives, and public policies*, Little Brown, Boston

Krimsky, S and Golding, D, 1992, *Social Theories of Risk*, Praeger, Westport, Connecticut

Lindblom, C, 1977, *Politics and markets: the world's political and economic systems*, Basic Books, New York

MacLean, D, 1980, *Risk and Consent: A Survey of Issues for Centralized Decision Making*, Center for Philosophy and Public Policy, College Park, Maryland

Maine, H S, 1861, *Ancient Law*, J Murray, London

Manne, A and Richels, R, 1992, *Buying greenhouse insurance: the economic costs of carbon dioxide emissions limits*, MIT Press, Cambridge

March, J G and Olsen, J P, 1989, *Rediscovering Institutions: The Organizational Basis of Politics*, Free Press, New York

Marriott, M, 1967, Hindu transactions: diversity without dualism, in: B Kapferer, ed, *Transaction and meaning,* Institute for the Study of Human Issues, Philadelphia, PA

Meadows, D H, Meadows, D L, Randers, J and Behrens III, W W, 1972, *The limits to growth*, Universe Books, New York

Meyer, A, 1995, The unequal use of the global commons, in: Katama (1995)

Nash, R, 1967, *Wilderness and the American Mind*, Yale University Press, New Haven

Nichols, A L, 1994, Risk-based priorities and environmental justice, in: A Finkel and D Golding, eds, *Worst things first: the debate over risk-based national environmental priorities*, Resources for the Future, Washington, DC

NRC, 1982, United States National Research Council, Committee on Risk and Decision Making 1982, *Risk and Decision Making: Perspectives and Research*, National Academy Press, Washington DC

NRC, 1996, United States National Research Council, *Understanding Risk in a Democratic Society*, National Academy of Sciences, Washington DC

Pahl-Wöstl, C, Jaeger, C C, Rayner, S, Schaer, C, van Asselt, M B A, Imboden, D, Vckovski A, 1998, Integrated Assessment of Climate Change and the Problem of Indeterminacy, in: P Cebon, H Davies, D Imboden, C C Jaeger, eds, *Climate and Environment in the Alpine Region – An Interdisciplinary View*, MIT Press, Cambridge, Massachusetts

Raiffa, H, 1982, *The art of science and negotiation*, Belknap Press, Cambridge, MA

Rawls, J, 1971, *A theory of justice*, Harvard University Press, Cambridge, MA

Rayner, S, 1982, The perception of time and space in egalitarian sects: a millennarian cosmology, in: M Douglas, ed, *Essays in the sociology of perception*, Routledge and Kegan Paul, London

Rayner, S, 1986, Politics of schism: routinisation and social control in the International Socialists/Socialist Workers Party, in: J Law, ed, *Power, action and belief: a new sociology of knowledge*, Routledge and Kegan Paul, London

Rayner, S, 1991, A cultural perspective on the structure and implementation of global environmental agreements, *Evaluation Review* **15**(1):75–102

Rayner, S, 1992, Governance and the Global Commons, Discussion Paper 8, Centre for the Study of Global Governance, London School of Economics

Rayner, S, 1995, A conceptual map of human values for climate change decision making, in: Katama (1995)

Rayner, S, and Cantor, R, 1987, How fair is safe enough: the cultural approach to technology choice, *Risk Analysis: An International Journal* **7**(1):3–9

Rayner, S and Malone, E L, eds, 1998, *Human choice and climate change*, 4 vols, Battelle Press, Columbus, OH

Robinson, J B, 1992, Risks, predictions, and other optical illusions: rethinking the use of science in social decision-making, *Policy Sciences* **25**:237–254

Rose, A, 1990, Reducing conflict in global warming policy: the potential of equity as a unifying principle, *Energy Policy* **18**:927–948

Rose, A and Stevens, B, 1993, The efficiency and equity of marketable permits for CO_2 emissions, *Resource and Energy Economics* **15**:117–46

Russell, B, 1946, *History of Western Philosophy*, George Allen, London

Shue, H, 1993, Subsistence emissions and luxury emissions, *Law and Policy* **15**(1):39–59

Shue, H, 1994, Avoidable necessity: global warming, international fairness, and alternative energy, in: I Shapiro and J W DeCena, eds, *Theory and practice*, New York University Press, New York

Smith, K, Swisher, J and Ahuja, D, 1993, Who pays to solve the problem and how much? in P Hayes and K Smith, eds, *The global greenhouse regime: who pays?* Earthscan/United Nations University Press, New York

Stone, C D, 1972, Should trees have standing: towards legal rights for natural objects, *Southern California Law Review* **45**:450

Thaler, R and Rosen, S, 1975, The value of saving a life, in: N Terleckyi, ed, *Household production and consumption*, National Bureau of Economic Research, New York

Thompson, M, 1979, *Rubbish theory: the creation and destruction of value*, Oxford University Press, Oxford

Thompson, M, 1987, Welche Gesellschaftsklassen sind potent genug, anderen ihre Zukunft aufzuoktroyieren? in L Burchardt, ed, *Design der Zukunft*, Dumont, Köln

Thompson, M and Rayner, S, 1998, Cultural discourses, in: S Rayner and E Malone, eds, *Human choice and climate change, Vol 1: the societal framework*, Battelle Press, Columbus, OH

Tönnies, F, 1887, *Gemeinschaft und Gesellschaft*, Wissenschaftlich, Darmstadt

Toman, M and Burtraw, D, 1991, Resolving equity issues: greenhouse gas negotiations, *Resources* **103**:10–13

UN (United Nations), 1992, *Framework Convention on Climate Change*, United Nations, New York
UN (United Nations), 1997, Kyoto Protocol to the United Nations Framework Convention on Climate Change, United Nations, New York
van Asselt, M B A, Rotmans, J, den Elzen, M, Hilderink, H, 1995, *Uncertainty in Integrated Assessment Modeling: A Cultural Perspective-Based Approach*, GLOBO Report Series 9 (no. 461502009). National Institute of Public Health and Environmental Protection (RIVM), Bilthoven, The Netherlands
Watson, R T, Zinyowera, M C and Moss, R H, eds, 1996, *Climate change 1995: impacts, adaptations and mitigation*, Cambridge University Press, Cambridge
Weiss, E B, 1989, *In fairness to future generations: international law, common patrimony, and intergenerational equity*, Transactional Publishers, Dobbs Ferny, NY
Welsch, H, 1993, A CO_2 agreement proposal with flexible quotas, *Energy Policy* **21**(7):748–756
Williamson, O, 1975, *Markets and hierarchies: analysis and antitrust implications*, Free Press, New York
Wirth, D A and Lashof, D A, 1990, Beyond Vienna and Montreal – multilateral agreements on greenhouse gases, *Ambio* **19**:305–310
Yamin, F, 1995, Principles of equity in international environmental agreements with special reference to the Climate Change Convention, in: Katama (1995)
Young, H P and Wolf, A, 1991, Global warming negotiations: does fairness count? *Brookings Review* **10**:2
Young, H P, 1993, *Equity in theory and practice*, Princeton University Press, Princeton, NJ

3 CLIMATE CHANGE AND MULTIPLE VIEWS OF FAIRNESS

Joanne Linnerooth-Bayer[1]

INTRODUCTION

At the recent Kyoto conference in December 1997 a consensus emerged on how much to reduce greenhouse gases in the short term and how to spread the costs of greenhouse gas abatement. The industrialized countries (Annex I countries) agreed to reduce their collective emissions by an average of about 5 per cent from 1990 levels in the period between 2008 and 2012. The principle of differentiated obligations was endorsed at Kyoto by agreeing to differences in the obligations on the part of industrialized countries and, more importantly, to the stipulation (by the gavel) that the developing countries (or the so-called South) will not 'evolve' towards commitments on emission reductions. In principle, the parties agreed to both joint implementation and emission trading, although the details remain to be worked out.

The Kyoto Protocol contains important principles of fairness for the present and future international response to the risks of climate change. Yet, the dissatisfaction on the part of the US and many other developed countries on the exclusion of the developing countries in a process towards commitments may jeopardize ratification of the protocol. Many open questions also remain about the form of joint implementation and emission trading that raise important issues of distributive fairness. After Kyoto, therefore, issues of equity will undoubtedly continue to be a central feature of any global action to mitigate global warming.

Even if the provisions set out in the protocol are realized by 2012, experts generally agree that there will be only very marginal reductions in the forecast climate warming of the next century (Bolin, 1998). This inevitably raises the question whether the benefits of the Kyoto emission-reduction commitments are worth their costs. One view holds that decisive global action to ameliorate destitution and impoverishment of persons alive today is a more effective use of global resources than greenhouse gas reduction, especially if the future

---

1 The author wishes to thank Anthony Patt, Thomas Sohelling and Michael Thompson, as well as two anonymous reviewers, for their very helpful comments on this chapter.

victims of global warming are better off than their ancestors are today and are thus able to adapt to a changed climate. Yet, as Rayner and Malone (1998) point out, such questions have hardly been posed in the Kyoto negotiating environment because of the hegemonic grip that the emissions reduction strategy has on the policy discourse.

This discourse has produced a wide range of proposals for sharing the costs of greenhouse gas (GHG) reductions between the North and the South. These proposals reflect not only the large number of possible criteria for distributing emission rights, but also the fundamental differences in the principles of fairness brought to the international policy debate. It might be argued that the Kyoto Protocol has made this discussion obsolete given the pragmatic political considerations that ultimately set the Kyoto commitments for greenhouse gas reduction, and that stray only slightly from the status quo. The important exception is the controversial principle placing responsibility for emission reductions fully on the industrialized countries. With predicted escalating emissions in the South, there is a gradual but significant redistribution of entitlements from the North to the South. However significant this redistribution is, nevertheless, it will hardly change the fact that per capita emissions on the part of the US and other industrialized countries have been, and will continue to be, far greater than the per capita emissions of the developing countries. The acceptability of this redistribution will be an important issue in the ratification of the Kyoto Protocol, especially in the US, as well as in future negotiations on its possible extension. In other words, issues of fair allocation will remain on the international negotiating agenda.

How can policy analysts and modellers engaged in climate research contribute to an international consensus in a fair way to allocate the costs of mitigating or adapting to climate warming? Chapter 3 of the report from the Intergovernmental Panel on Climate Change (IPCC, 1995:52) suggests that insights from local experience on dividing and allocating environmental burdens can be useful for this task:

> *...there is limitless practical experience on how countries handle distribution issues. It is an interesting question whether and how any of this analysis and experience should be brought to bear on the issue of climate change. Interestingly, however, none of the climate change literature at present appears to attempt this.*

Taking up the challenge of the IPCC, this chapter draws on national experience in distributing another type of environmental burden, namely that of hazardous waste disposal. Since any disposal strategy inevitably imposes risks and costs on selected communities, regions or countries, a core issue is how societies impose these risks for the overall benefit of the society producing the wastes. Likewise, how does the global community impose on countries the burden of reducing GHG emissions for the overall benefit of the global community? Both questions involve issues of rich and poor, generators and non-generators, winners and losers, and contending views of what is fair *within* countries as well as between countries.

This discussion examines two questions of fairness in the climate change debate. Should nations continue along the path begun at Kyoto by allocating extensive resources to greenhouse gas abatement over the next decades, keeping in mind the competing demands for these resources? And, given a resolve to abate GHG emissions, what is a fair way to allocate the costs? This chapter discusses the multiple moral discourses on each of these issues, and argues that an understanding and respect of the plural views of fairness is important in crafting an agreement that can be implemented. It also examines whether views of fairness can be related to recent theories concerning the social construction of environmental risks. Throughout the discussion, this chapter illuminates the arguments with examples from an IIASA study on distributing the burdens of hazardous waste disposal in Austria, which showed that views on fairness are empirically verifiable and related to both interests and culturally determined values (Linnerooth-Bayer and Fitzgerald, 1994).

MULTIPLE VIEWS OF FAIRNESS

Experience with regard to siting hazardous waste facilities and other issues of local justice show that there is no single, universally valid set of moral principles that can inspire effective personal, institutional or national commitments to a fair distribution of environmental risk burdens (Linnerooth-Bayer and Fitzgerald, 1994). Competing conceptions of what is good and bad – acceptable and unacceptable – characterize most policy issues, with an associated plurality of values and interests regarding distributive fairness. The siting experience shows that discourses on fairness are not only tied to national cultural differences or interests – the North and the South – but that cultural cleavages can be identified within all communities. These cleavages are reflected in the diversity of political views and worldviews that coexist within each polity. As anthropologists have pointed out, the variations in political attitudes and values within countries are at least as great as between countries (Thompson et al, 1990). Patt (1997), for instance, has shown how economists and ecologists have offered differing opinions about the seriousness of climate change, and these differences are related to the different values brought to the analyses.

A leading expert on distributive justice at the local level, Elster (1992:15) argues that these competing conceptions of fairness are context dependent, so much so that it is difficult to generalize:

> *I do not think the study of local justice will ever yield much by way of robust generalizations...local justice is above all a very messy business. To a large extent it is made up of compromises, exceptions, and idiosyncratic features that can be understood only by reference to historical accidents.*

At the other extreme from Elster, and based in large part on studies from child psychology, Wilson (1992) argues that fairness is part of a universal moral sense; it is, therefore, neither environmentally determined nor culturally relative. Young (1994) takes a middle view and notes that although generalizations are difficult, similar concepts do continue to reassert themselves. Douglas (1985) and other cultural theorists (see Thompson et al, 1990) suggest how and why these reassertions take place. According to cultural theory, views of fairness are not ad hoc, but they reflect predictable cultural biases and worldviews (which also contribute to determining interests). These theorists identify how divergent moral principles bias the social construction of fairness (Rayner and Cantor, 1987). The central idea of this theory is that in order to maintain solidarity, different forms of social organization (cultural groupings) generate predictable arguments concerning fair process and outcome, including procedures for allocating responsibility, for self-justification or for calling others to account. If this is correct, then it is important to understand and recognize the contending ideas of fairness and how they vary across the social, political and cultural fabric.

In an attempt to explain the different positions on the climate change issue taken by the European Union, Japan and the US, Johnson (1995) has hypothesized that the dominating national positions are related to worldviews, and that these differences go beyond national interest. Whether such an understanding can be useful for designing agreements on allocating the social burdens of mitigating and adapting to climate change depends importantly on the extent to which distributive justice plays a role in international relations. In this volume, Victor argues that international agreements are motivated first and foremost by national interest or willingness to pay for the expected benefits from the agreement. Victor tacitly rules out re-distributive effects as a benefit for which nations are willing to pay. While the historical record suggests that willingness to pay on the part of nation states for international commitments is, for the most part, bereft of humanitarian and moral commitments to global redistribution, and the US negotiating position at Kyoto appears to support this observation, this may be changing. Demands for redistribution may move more to the forefront as global democratization forces societies to open their political systems, thus empowering egalitarian and other groups with global distributive interests.

CLIMATE WARMING: HOW MUCH IS FAIR?

In a provocative argument, Schelling (1998) questions whether a global policy of reducing GHG emissions is the most cost-effective way to improve the welfare of the world's poor. The beneficiaries of climate-change abatement policies will overwhelmingly be in the developing countries. Nine-tenths of the population is expected to be in these countries in 75 years' time (when the benefits accrue). Since any costs of mitigating climate change during the coming decades will almost certainly be borne by the wealthy countries, climate abatement should be viewed, according to Schelling, as a huge foreign-aid programme for the future generations of today's developing countries.

The issue Schelling (1998:13) raises is whether this is the best use of foreign-aid funds:

> *Is there something escapist about discussing 2 per cent of GNP to be invested in the welfare of future generations, when we do nothing like that for their contemporary ancestors, a third of whom are so undernourished that even a case of measles can kill them?*

As Schelling and others promoting this utilitarian argument acknowledge, it rests on a number of 'ifs'. If natural and man-made capital are substitutable and climate change adaptation is possible without large social costs, for example, from catastrophic consequences, and if a stable climate does not have value divorced from its impact on human beings, and if we accept intergenerational consumer sovereignty, then GHG abatement is a questionable policy in terms of improving the welfare of the world's poor. Would the developed countries not get more from their investment in terms of overall welfare for the people living in the developing countries if they invested in *current* programmes that benefit the destitute and vulnerable? Is this not what those living in the developing world would choose if they were given the money? This argument gains appeal when one considers that many of the future beneficiaries of climate change policy will, at least under commonly held assumptions, be far better off than many in the developing world today and thus better able to cope with a changed climate. Moreover, if the investments are made wisely, they can build the necessary infrastructure such that future generations can cope even better. The contemplated trillion or more dollars for greenhouse gas reduction, therefore, may be a grossly inefficient investment to improve the welfare of the poor people of the world, across and between generations.

This is more than an argument for cost effectiveness or efficiency in helping the poor people of the world. It is a view of fair allocation. Allocating to improve the welfare of future generations means depriving those today who are even more in need. To a global utilitarian, whose ideal is to bring about the best state of affairs in terms of *overall* welfare or happiness, this is unfair, even immoral. In this sense, efficiency is a moral argument.

The utilitarian argument is consequentialist. The utility arising from our actions is what is important. One well-known objection to utilitarianism is that it does not matter *whose* utility is being increased or reduced; therefore, a poor person's welfare can justifiably be reduced to increase that of a wealthy person. One cannot, however, object to this utilitarian argument on the grounds that it discriminates against the disadvantaged in order to increase utility overall. Schelling takes a Rawlsian position as his starting point by assuming that the objective is to increase the welfare of those in the poor countries. He examines how we can do this most effectively.

The welfare economist might also view the trillion dollar investment as unfair. Welfare economics is a branch of utilitarianism. Yet, economists concede that interpersonal – and thus intergenerational – comparisons of utility are not possible. Without intergenerational utility comparisons, one cannot substantiate the utilitarian argument that allocating to development aid will increase current utility to this generation more than it will detract from the utility of the next. To unequivocally state that there is an overall welfare gain between generations, the economist would require that *both* generations come out ahead. This Pareto claim (in contrast to the utilitarian claim) can, indeed, be made if investing in the economies of the developing countries reduces current poverty *and* creates the infrastructure, such that future generations can better cope with a changed climate. To the extent that those living in year 2075 are better off with the economic growth (and reduced vulnerability) from current investments than in GHG reductions, the strategy of investing in the economies of the poor countries is Pareto superior to abating climate change.

The Pareto criterion is a central concept of welfare economics. According to this criterion, state of the world A is socially preferable to state of the world B if at least one person prefers A to B and no one prefers B to A. It is important to appreciate the subtle difference between this Pareto form of utilitarianism that arises from individualized choices, what Rayner has referred to as 'market utilitarianism', and the classical utilitarianism of Bentham with its top-down inventorying of the stocks of 'happiness' or utility. Classical utilitarianism requires a judgement concerning the welfare of the aggregate, and this judgement is divorced from any notion of individual rights or choices. This empowers a state hierarchy to act in the interests of the social good. As Rayner points out (see Chapter 2), the development of the behavioural paradigm in social science has created the possibility of utilitarian decision making and provided a moral basis for hierarchical organization. The condition imposed on this paradigm by economists – that each individual must judge if he or she is better off, alternatively – decreases the power of the state by invalidating top-down judgements of social welfare. Market utilitarianism, in sharp contrast to classical utilitarianism, is thus a far more individualistic and less hierarchical concept.

Would not the developing countries, themselves, choose development aid over GHG abatement if they were given the choice? This question, which Schelling asks, establishes his moral argument in the realm of market utilitarianism. It is inconceivable to most economists that a Pareto improvement – where all concerned feel themselves to be better off – is not viewed as a fair outcome. If both generations are better off from present day investments, can anyone still reasonably argue that a policy of GHG emissions abatement is more fair than one of helping today's poor?

Of course, whether current day investment in economic infrastructure in the developing countries is a Pareto superior policy to GHG abatement is contingent upon many controversial assumptions. The point of this discussion is not to examine whether these assumptions hold. But if they do hold, might there still be legitimate claims for GHG abatement on the grounds of fairness? In other words, is the efficiency argument underlying market utilitarianism the only legitimate claim to a fair outcome? This question will be examined in the next section in the context of hazardous waste export policy.

VIEWS OF FAIRNESS AND EXPORTING HAZARDOUS WASTE

One of the more controversial international issues has and continues to be waste exports, or transporting hazardous wastes for disposal from rich to poor countries. At the core of the debate is whether the developed world is morally justified in shipping its wastes and environmental risks to countries that give their informed consent and that have adequate infrastructure to deal appropriately with the wastes. In an early and controversial World Bank memorandum, Summers argued that with adequate compensation and the provision of suitable infrastructure, both the exporting and importing country can be made better off with a flourishing trade in wastes (*The Guardian*, 1992). Indeed, the compensation paid for the waste exports might even result in an outcome that reduces overall environmental hazards and risks to the importing country. From a market–utilitarian view, this trade is an opportunity to increase welfare all the way around. Indeed, with knowledge and consent on the part of the importing country, this deal is a Pareto superior proposition. Like the market–utilitarian logic underlying the plea for development aid over GHG abatement, the logic supporting this policy, based on the theory of comparative advantage in waste disposal, is unassailable. Yet, legislation to restrict exports of hazardous wastes, especially to poor countries, has been motivated by the indignation, even moral outrage, that these exports arouse. Understanding this outrage gives important insights on the multiple views of fairness.

The Pareto proposition in the context of waste exports was tested in a survey given to the Austrian public (Linnerooth-Bayer and Davy, 1994). Like many other countries, Austria faces an impasse in siting hazardous waste facilities, and its poorer neighbours to the east have expressed interest in hosting a state-of-the-art facility to take Austria's wastes. Randomly selected members of the Austrian public were given the following question:

> *A neighbouring country with large-scale environmental problems and few resources for dealing with them has offered to host a state-of-the-art hazardous waste disposal facility for compensation. It can use this compensation to deal with its very serious problems such as improving its air and water quality and reducing its high rate of cancer in children. Should this offer be accepted?*

In response, 84 per cent of the respondents answered 'no' to this question. That is, they rejected a Pareto superior policy or one where all those affected considered the policy to make them better off. To a welfare economist, this overwhelming rejection of a Pareto superior deal is remarkable, and an understanding of the underlying rationale might shed light on the apparent rejection of similarly motivated arguments in the climate change debate.

One explanation for the negative response to this Pareto deal may lie in the concept of *responsibility*. On another question, responses to the Austrian questionnaire emphasized the importance placed on the notion that those producing wastes, whether a country or an industry, should directly bear the burdens or responsibility for their disposal in 'their own backyard' (Linnerooth-Bayer and Fitzgerald, 1994). The importance of a region or country taking responsibility for its own wastes was also supported by the results of a survey carried out in Switzerland, which showed that outright compensation is unacceptable to a large number of people because it detracts from the motivation of accepting the facility as a social duty or responsibility (Frey et al, 1995; see also Lidskog and Elander, 1992; Kemp, 1990; Easterling, 1992).

These studies show that Aristotle's 'equity principle', which holds that social goods (and bads) should be allocated in strict proportion to each claimant's contribution to the good, or responsibility for the bad (Young, 1994), may be relevant in understanding opposition to market–utilitarian principles. Since the industrialized countries have historically contributed more to GHG emissions, many view it as only fair that the polluters take primary responsibility for the climate change problem (IPCC, 1995). Alternatively, for the most part, the North does not feel directly responsible for today's poverty in the South, and foreign aid is usually viewed as charity instead of the North's responsibility. Since the North has contributed the most to global climate warming, and generally feels less responsible for current poverty in the South, the equity principle would privilege the reduction of GHGs over direct transfers to the poor. Those holding this view – that the

North has a moral imperative or global responsibility for reducing the future suffering in the world resulting from climate change – would reject the consequential logic inherent in the market–utilitarian argument.

VIEWS OF FAIRNESS AND SOCIAL ORGANIZATION

Both claims to fairness, the market–utilitarian appeal for the efficient allocation of resources to increase the welfare of the world's poor in this and the next generation, and the counter appeal for taking responsibility for GHG emissions (or wastes), are legitimate in the sense of promoting laudable social objectives. Yet, as seen in the hazardous waste case, they can lead to social conflict and competing public policies. Most importantly, an implementable policy will likely require a compromise between these and other notions of a fair outcome.

How can the analyst contribute to this compromise? If views on a fair outcome are essentially ad hoc and contingent on context, as Elster claims, there is little role for the analyst in the international consensus-building processes. On the other hand, if there is a consistency in the views of the policy actors, then there is a basis for empirical documentation that can be useful in the design of consensual policies.

Institutional analysis suggests that there is such a consistency, that views of fairness are constructed consistently with principles that are inherent in particular forms of social organization and social relations. According to Rayner (1994), views on fairness cannot be separated from ideas about community and social organization, or the need to establish shared values for the conduct of community procedures and the distribution of rights, goods and burdens. Views of fairness are maintained by reducing the scope of awareness, and some perspectives or values are selected as ideologies (March and Olsen, 1989). According to Douglas (1985), whatever principles of fairness may be legitimately argued, social organizations will select for attention those that maintain social solidarity. Following on from Douglas's proposition, cultural theorists have identified how this solidarity or cultural grouping is related to divergent worldviews. The question we examine is the extent to which these worldviews bias the social construction of fairness and the extent to which an understanding of worldviews can aid in the empirical validation of views of fairness.

What cultural theory assumes is, firstly, that cultural bias (in the form of different worldviews) is everywhere and always present (note that this is not national or regional culture) and, secondly, that there are a limited number of viable forms of social solidarity and organization: hierarchy, individualism, egalitarianism, and fatalism (Thompson et al, 1990). Hierarchical organization is characterized by positional authority, inequality and procedural rationality and, therefore, stands in sharp contrast to individualistic forms of social organization with their emphasis on personal initiatives and rights. Cultural theorists have added a third active worldview, the egalitarian. Egalitarian discourse inevitably emphasizes social equality and moral rightness, and generally takes a more holistic approach to policy issues. Looking through the lens of the cultural theorist, we would expect that appeals to responsibility (in response to the consequentialist reasoning of the market utilitarian) would appear in the discourse of all three of the so-called active cultural groups – hierarchy, egalitarianism, and individualism – but in different ways and for different reasons.

Utilitarian fairness

In hierarchy, fairness and distributive issues are settled by administrative determination based on such considerations as the rank of the recipients, their needs, their contributions and their responsibilities (Rayner, 1994). The normative imperative of the classical utilitarians to increase overall happiness is consistent with hierarchical fairness, but it is the administration that determines who is in need, who has contributed and who is most deserving of welfare transfers. Alternatively, market utilitarians take the overall welfare as defined by the free choice of those affected by public policies to be a key aspect of fairness. Hierarchical discourse makes little use of this substantive logic in favour of procedures that promote a sense of common moral purpose and that, above all, are easily justified. Of course, hierarchical determinations of welfare are not divorced from the welfare that the citizens, themselves, attribute to various policy choices. As Rayner states in this book, the greatest happiness to the greatest number (has become) the key to retaining elected office (see Chapter 2). In contrast to monarchies, democratic state administrators are obliged to consider the preferences of their constituencies and, in this way, classical utilitarianism takes on many aspects of market utilitarianism. As an important manifestation, in the US (but not in most European governments) individualistic concepts such as willingness to pay in calculating costs and benefits are, albeit often reluctantly, taken into account in the considerations of public policy-makers.

Kukathas and Pettit (1990:95) speak of communitarians who reject the self-determinism and the anti-government views of the liberals in favour of the power of structural roles and responsibilities. Responsibilities are implicit in the communitarian goal of creating a common good, and arguments of state responsibility enhance the role of the state bureaucracy. Especially in Europe, which makes greater use of classical utilitarian arguments in public policy justifications, we find hierarchical appeals to global responsibility to dominate over appeals for global efficiency. The discourse often calls for stewardship or the care and preservation of the global atmospheric commons for current and future generations. Such an appeal presupposes a concept of global management or even, some might argue, global governance.

Returning to the hazardous waste analogy, it is not surprising that governments throughout Europe have adopted stringent legislation against exporting wastes out of the country in favour of a larger state role in their management. As a case in point, Linnerooth-Bayer and Davy (1994) have shown that the Austrian government has explicitly legitimized this role with popular appeals to responsible waste management; market–utilitarian arguments for efficiency have had little place in the Austrian government's rhetoric. To the state hierarchy, the substantive rationality inherent in claims for increasing mutual welfare across borders by respecting the choices of the exporters and importers is not as appealing as arguments for establishing procedures that promote a sense of moral rightness and cohesion in the society. Procedural rationality trumps over market–utilitarian, consequential rationality. Alternatively, in the US, with its more individualistic cultural heritage, there has been more reluctance to enter agreements limiting international trade in wastes (Linnerooth-Bayer and Löfstedt, 1996).

Market–utilitarian fairness

Individualistic forms of social organization are characterized by a more substantive rationality with appeals to personal initiative, freedom of choice and individual rights (Schwarz and Thompson, 1990). In contrast to most other theories of justice, libertarian theories hold rights to be supreme, and responsibility takes the form of protecting and honouring these rights (Kymlicka 1990:275). There is no moral imperative in the individualist ethic to transfer large sums of money in development aid to the South. On the other hand, if the North imposes hardships on the South from GHG emissions, there may be a moral obligation or responsibility to rectify this inequity. However, this depends importantly on how international rights, liabilities and obligations are interpreted.

The rights of the victim states as set out by the Stockholm and other international law principles of responsibility would figure importantly in the individualistic ethic; however, the common law practice suggests that these rights do not yet exist in international practice. In some cases, states have accepted liability for transborder damages – for example, the Soviets in the case of the Cosmos 954 – but more often states have denied responsibility (Linnerooth-Bayer and Loefstedt, 1996). After Chernobyl, for instance, German and other states' requests for compensation were turned down by the Soviets. According to Zemanek (1992:188), there is not a 'widespread, constant, and consistent practice of States necessary for generating a rule of customary law', and 'we do not yet have a general acceptance among states of the proposition that states are either responsible or liable for damage or harm caused to the environment of other states or of the commons'. State responsibility for transboundary damage has thus not been unequivocally established in the 'customs of nations' or in customary law (see Bodansky, 1995).

Fairness to the individualist – according to Young (1994:1) – is not founded on a universal sense of morality but is shaped in part by principle and in part by precedent. There is little precedent supporting the notion that sovereign states either reduce their transborder pollutants (or risks) or compensate the victims. Based on precedent, the logic of imposing an obligation on the part of the developed countries to reduce their emissions beyond what they would consider to be in their interests is hard to justify. To the extent that national interest is demonstrated by public choices, which is a decidedly individualistic claim, it would appear that there is little interest on the part of the North in aiding the South. Current generation, North–South transfers are nowhere near on the scale of the expenditures envisioned for climate change abatement. Since the publics of the wealthy countries appear not to be willing to pay much to increase the welfare of the poor in the developing countries, the individualist ethic would suggest that there is no apparent reason (outside legally assigned obligations which have not been established in international law) to reduce GHGs for this same purpose. Nor is there a convincing argument to increase foreign aid to the poor countries in lieu of reducing GHGs.

In sum, the utilitarian argument for efficiently allocating resources to the poor of the world could be construed by the market utilitarian to mean that the North should abstain from engaging in massive aid programmes *and* in GHG emission abatement. The rejection of any moral imperative to aid the South is rooted in moral claims based on the respect of individual choice and rights-based approaches to public policy. There is no precedent of collective choice for abating transboundary burdens or for massive distributions to the poor of the South. Besides substantial evidence of public preferences for redistributing from the North to the South, there is otherwise little obligation or responsibility for the North to engage in GHG emission abatement unless, of course, these abatements are in the North's own interests.

This claim to fairness will, no doubt, figure importantly in the debate on the ratification of the Kyoto Protocol. Already in the US, arguments that show that meeting the treaty will be in the overall benefits of the country are an important part of the political debate. Already some US economists are arguing that the Kyoto Protocol could benefit the economy in light of the country's competitive advantage in the race to reliable solar power, gasoline alternatives based on agriculture, zero-emission 'fuel cells', and similar technology (Easterbrook, 1997). As put by Tumulty in *Time* magazine (1997:13),

> *...the campaign to convince the country that the treaty's targets are reasonable and relatively painless will begin next month...Its success in the end will depend on expanding America's environmental constituency: will the soccer moms give up their sport-utility vehicles for a future of fresh air?*

Egalitarian fairness

Many different types of moral claims support the notion of responsibility over the utilitarian's consequential rationality. Responsibility factors strongly into Rawls's (1971) original position. The moral reciprocity in the veil of ignorance forces individuals to treat others as they would want to be treated themselves, making responsibility to fellow humans an intricate functional property of Rawls's justice scheme (Kymlicka 1990:278, as discussed in Van Well, 1997:7). In opposition to Rawls, the feminist ethic suggests that people's moral actions can respond to the complexity of a particular situation without the need to explicitly assert moral principles (Van Well, 1997). Accordingly, methods of reasoning emphasize the specific rather than the general, and responsibilities and relationships take precedence over rights and efficiency (Kymlicka 1990:265).

Egalitarian rhetoric generally takes a more holistic approach to environmental issues. The need to site hazardous waste facilities, for instance, is suspect when one considers the possibilities for reducing or even eliminating hazardous wastes. Therefore, it is not a question of dealing with the wastes at home or abroad, but of radically changing production practices to eliminate wastes. Likewise, the choice of allocating between development aid and GHG abatement is a misguided choice. Both are necessary, especially given the North's historical role in the exploitation of the South. Arguments of efficient allocation given scarce resources are often viewed in egalitarian discourse as a thinly disguised manoeuvre to justify the inequitable status quo and to avoid the ultimate task of radically changing the way we live.

The view of many environmental groups has, thus, understandably been negative towards the incremental progress on abating climate warming. Hermach of the Sierra Club, for instance, referred to Gore's post-Kyoto announcement of a new partnership between the electronics industry and the US Environmental Protection Agency as 'a long way from the agenda Gore laid out in his 1992 enviro-treatise, *Earth in the Balance*: that an overhauling of the economy is necessary to avert disaster' (Boal, 1997:3).

But how can egalitarian discourse justify arguments against a Pareto proposition that guarantees *both* generations an improvement in their welfare? In an earlier section, we argued that *responsibility* is a key concept in moral discourses that argue against Pareto efficient policies. The next section reexamines this question and the multiple views of fair allocation, calling again on the empirical work on hazardous waste disposal.

VIEWS OF FAIRNESS AND SITING HAZARDOUS WASTE FACILITIES

The impasse throughout the industrialized world in siting hazardous waste facilities can be attributed, in part, to the failure of most siting initiatives to take adequate account of the diverse views held by the public and other stakeholders on a fair siting process and outcome (Linnerooth-Bayer and Fitzgerald, 1994). In most traditional siting processes, central governments have acted as trustees for the interests of the larger society by relying on experts for technical assessments of risk and other site conditions. These hierarchical, cohesive approaches appeal to those who trust their government and its network of experts, but they are opposed by those who desire more personal influence and responsibility for the siting decision. Many observers argue that if genuine cooperation on the part of the public is desired, the control and decision authority should rest more firmly on the affected communities and their citizens (O'Hare et al, 1983; Mitchell et al, 1986; Kunreuther et al, 1987).

Taking the individualistic view, the solution lies in changing the balance between perceived local risks and benefits, not by administrative determination, but by granting the citizens of the prospective host communities the right to negotiate them to their advantage. Supporters of this voluntary market approach argue that given collective consent mechanisms such as referenda, residents will voluntarily enter into siting contracts (Kunreuther et al, 1993; Gerrard, 1994). Advocates of this approach point to the fact that it guarantees a Pareto improvement: one in which the siting decision (with compensation) is preferred over the status quo by all concerned (O'Hare et al, 1983; Kunreuther et al, 1987; Laws and Susskind, 1991). It follows that advocates of the market approach view *compensation* as the key to community consent and reaching mutually agreed solutions to the siting impasse.

However, voluntary market approaches to siting facilities are vehemently opposed by those who believe that they will further widen the gulf between the rich and the poor by siting facilities in poor communities (Bullard, 1993). The reason is that the poor, because of their economic circumstances, will likely have a lower 'reservation price' for lost amenity and exposure to health risks than will the wealthy. Not only might negotiated compensation lead to placing facilities in poor communities, egalitarian discourse faults compensation for not assigning significance to values such as the sacredness of life. Ample evidence now exists that many citizens are reluctant to negotiate a price for accepting a hazardous facility (Portney, 1985), and many view compensation as nothing less than a bribe.

Egalitarian discourse is also critical of solutions that do not address the broader social issues responsible for the waste problem, pointing out that with more recycling and less waste the facility might not be needed at all. In the words of Thompson (1994:169):

> *...those who favour the egalitarian solution, and who wish to radically transform the production and consumption system that is all the time generating these hazardous wastes, will see both markets and hierarchies as inevitably perpetuating rich–poor divides, on the one hand, and exploitative and unsustainable technologies, on the other.*

The survey of Austrian opinion demonstrates these different views of a fair process and outcome for siting hazardous waste facilities (Linnerooth-Bayer and Fitzgerald, 1996). Many Austrians appeared to support the traditional, hierarchical structures with legitimacy rooted in technical expertise; others held strong egalitarian views for spreading the burdens as evenly as possible through geographical equity. Many were concerned about distributive criteria, especially contribution to the waste problem and community vulnerability; and still others were strongly individualistic and supported local rights and market mechanisms for allocating the hazardous waste burden. The survey also illustrated both the support and the moral indignation that accompany offers of compensation. When respondents were asked how they would view compensation of US$3,000 per year for accepting a hazardous waste facility near their community, *39 per cent* thought it would make the facility more acceptable to them. However, *31 per cent* felt it would make little difference and *29 per cent* actually thought it would make the facility less acceptable!

This finding has little to do with cultural differences between countries, or between the North and South, but shows the multiple perspectives on what is fair in a single community. It does not mean that these perspectives are not related to interests, but it does suggest that interests cannot be construed as simple economic advantage. This and similar findings of local justice may also help to explain the lack of any universally agreed moral principles for abating greenhouse gases and the opposition to any one moral discourse. Might, for instance, the argument for endowing the future generation of the developing world with a better economy and infrastructure so that they can better adapt to a changed climate be viewed as a form of compensation for the risks of climate change? Might Schelling's argument be construed as bribing future generations into accepting the risks of climate change in exchange for economic prosperity and reduced vulnerability? In the egalitarian view, these risks are extremely serious, even potentially cataclysmic (Rayner, 1997). Compensating for horrific risks is an irresponsible strategy.

Moreover, egalitarian claims to fairness do not necessarily value all-around improvements if they serve to increase inequality. MacLean (1993) has put this succinctly. If the initial endowment between the rich country and the poor is, say, (100:30), the egalitarian may prefer a move to (65:25) over a move to (195:35). In other words, egalitarian fairness might in some contexts advocate less for everybody if it is distributed more uniformly. This is in direct violation of the Pareto criterion. In contrast to utilitarian consequentialism, egalitarian and other distributive justice theories concern themselves with the perceived fairness of the distribution, per se. The Austrian survey yielded evidence that this phenomenon is empirically verifiable.

This concern for strict equality, even equality of responsibility, over efficiency is apparent in the discourse of groups in the developing countries as well as environmental groups who oppose market-based solutions. Economists have been baffled, for instance, by objections in the developing countries to joint implementation, where industrialized countries invest in reducing emissions in the developing countries (the investments would bring more emission reduction for the money spent than in the industrialized countries). There are many arguments against this policy; the most egalitarian is that joint implementation allows industrialized countries to continue increasing their own emissions. This will perpetuate, even exacerbate, global inequalities in per capita emissions (Parikh, 1995). A similar concern about Pareto superior policies that widen the gap between North–South emissions among the countries of the world was voiced at the Kyoto negotiations in the developing country opposition to the concept of emission trading. Some commentators have pointed out that, for instance, India's objection to emission trading was to flag the need for convergence of country per capita emissions (World Energy Council, 1997:11).

This does not mean that joint implementation or emission trading will not eventually be adopted as a way of promoting a more efficient reduction of GHG emissions (they have been agreed in principle in the Kyoto Protocol); however, it means that they will not be enthusiastically embraced by all the policy actors. Experts estimate that trading credits to emit carbon gases with the so-called 'umbrella' group countries (at this writing, the US, Canada, Australia, New Zealand and Russia) could cut costs of complying with the global warming treaty by up to 75 per cent compared with having no trading system (Yellen, 1998). A large part of this reduction may, however, come from the so-called problem of 'hot air' since Kyoto gave Russia and other transitional countries credits for greenhouse reductions already achieved by the closing of antiquated Soviet factories. These credits can then be bought by the US and other developed countries. The continual urging by the US for flexible, market-based mechanisms, such as emissions credits trading, was at odds with the European Union, which was concerned that the plan of shifting emissions reductions to other countries will produce big loopholes that could undermine the treaty.

It will be important to frame the discussion and shape the discourse in terms that will appeal to the diverse interests. For example, some US environmentalists supported the sulphur dioxide (SO_2) tradeable permits once they realized that groups could buy the permits as a way of reducing emissions altogether. At Kyoto, the term of joint implementation was dropped and replaced by a 'clean development mechanism', a more acceptable term for egalitarian discourse.

Allocating the Burden of Reducing GHGs

By setting emission reduction commitments (and exempting the South altogether), the Kyoto Protocol implicitly established emission rights and the distribution of the emission reduction burden. Future negotiations, and especially any considerations to bring in those countries not listed in Annex I, will require changes in the current emissions regime. As Young (1994:3) notes, property rights are constantly being created, destroyed, revised and reassigned by society. The plurality of arguments for distributing emission rights, which one finds in the climate change discourse, reconfirms our theme of contending ideas of fair distribution. The cultural theory lens, along with the literature on distributive justice, suggests a very simple taxonomy of fairness principles for these arguments.

Equality

As anyone who has divided a cake at a birthday party knows, an intuitive and appealing concept of fairness is *equality*. There are many variations of equality in the GHG emission rights discourse, the most obvious being equal per capita entitlements (Grubb, 1990; Bertram, 1992; Epstein and Gupta, 1990; and Agarwal and Narain, 1991). Although proponents usually suggest an accompanying disincentive for population growth, per capita entitlements are a form of division according to equal shares. Egalitarians do not always, however, embrace *strict* equality but often argue in favour of redressing *past* inequities. As a case in point, some groups have expressed a preference to locate hazardous waste facilities in wealthy neighbourhoods since the wealthy, by way of their consumption, have contributed more to the hazardous waste problem (Linnerooth-Bayer and Fitzgerald, 1996). In the emissions rights discourse, many commentators also argue for disadvantaging the wealthy countries on the basis of their historical contribution (see Grübler and Fujii, 1991). This idea of fairness is not based upon fault, blame or compensation, but upon an egalitarian principle of access (Rayner, 1997).

An objection, in principle, to this egalitarian claim that fits comfortably with individualistic discourse concerns fault, liability and blame. If past emitters were ignorant of the consequences of their actions, should they be required to pay damages in terms of reduced emission rights? This notion of strict liability also finds its analogy in the hazardous waste case. In the US, those who legally disposed of hazardous wastes in the past and without knowledge of the consequences are now held liable for their cleanup and damages. This legislation finds strong support in the ranks of egalitarians, but is viewed as unfair by many others (Baron et al, 1993).

Deviations from equality

Much of the philosophical literature on fair allocation has been concerned with justifying deviations from equality. The concept of differentiated obligations found in the First Principle, Article 3 of the Climate Change Framework Convention, is an appeal to deviations from strict equality in the share of the burden. Not surprisingly, we find incompatible and competing arguments for allocation in terms of the varied justifications for inequality.

As summarized by Rescher (1966), one can think of four fundamental concepts that can be appealed to in arguing for deviations from equality: need, endowment, merit and contribution. Distribution to those who have the greatest *need* is the basis of such commonplace institutions as rules for public housing, food stamps and need-based financial assistance for higher education. *Endowment* of talent, ability, or property, and *merit* are the fundamental notions behind meritocracy, most scholarships and interest income. Finally, reward according to effort and productivity, or *contribution*, is the foundation of commission income and school grading systems.

We find frequent reference to these concepts in arguments for deviating from equal shares in the climate debate. *Need* is appealed to in arguments for allocating according to land size, or allocating such that there is a comparable burden, or allocating in proportion to GNP (Cline, 1992). The US negotiators at Kyoto justified their arguments for a smaller commitment in GHG reductions than Europe's commitment on the grounds that the US is expecting greater economic growth and thus will need higher emissions. *Contribution* is appealed to in placing responsibility on the industrialized countries and, more strongly, in arguing for punishing big emitters by taking into account historical emissions. It was also appealed to at the Kyoto negotiations by the US to put eventual responsibility on the developing countries. *Endowment* (or capabilities) is also appealed to in arguments for requiring the wealthy countries to bear the brunt of the burden since they are better endowed for this task. The endowment claim is explicitly stated in the Framework Convention: 'the developed country parties should take the lead in combating climate change and the adverse effects thereof'. *Merit* is appealed to in rejecting, for example, equal percentage or flat reductions since this would punish those countries that have already reduced their GHG emissions and thus merit exemption.

It was the contention between claims of *contribution* and *endowment* that played a central role in the Kyoto debate. The US delegation, and later the New Zealand delegation, argued for meaningful participation of key developing countries and a process for eventually setting commitments on their part. Keeping in mind the large differences in per capita emissions, today and historically, the Group of 77 and China immediately objected to any proposal for a post-Kyoto evolutionary process, and they rejected any link between technological and financial assistance and commitments. The line taken was a demand for transfers from the North to the South on non-commercial, preferential terms and resentment at the emphasis placed by many developed countries on the market system and the private sector (World Energy Council, 1997).

Preferences and rights

Individualistic discourse, alternatively, shows a much greater emphasis on the private sector and on preferences, rights, liabilities (responsibilities) and incentives (Schwarz and Thompson, 1990). If the rich countries have no special responsibility for abating climate change damages (and we discussed above the ambiguity in this responsibility depending on international law principles and international law precedent), an appropriate principle for contributing to the costs of abatement would depend on individual or country preferences – what the countries would be willing to pay (WTP) in light of what they perceive as their potential damages from climate change. Or, as Dorfman (as reported in IPCC, 1995:51) shows, willingness to pay for a common property resource should be based on how much a country values a reduction in worldwide aggregate emissions. According to this principle, contributions will be determined by a combination of ability to pay and perceived benefits. Summing up WTP for climate change abatement across the globe now and in the future (appropriately discounted) should, in theory, determine the aggregate value of reducing GHGs.

The controversy over the principle of aggregate willingness to pay to value the benefits of reducing climate change illustrates our theme of conflicting values and perspectives. The notion of allocating resources according to individual preferences, and aggregating these preferences, is perfectly valid from a market–utilitarian and welfare–economic view. Only if the people of the world (and with an appropriate discount factor for the future people of the world) value the consequences of reducing climate change above the costs, should the trillion dollar investment in GHG abatement be taken. If the benefits are more than the costs, we can correct the distributive inequities with compensation and come out ahead.

Objections to the notion that preferences should count in valuing the damages from climate change were dramatically demonstrated by the response to the social cost chapter of the *Second Assessment Report* of the IPCC Working Group II (Pearce et al, 1996). Since the publics in poor countries cannot – and are not willing to – pay as much for abating the damages of climate change as the publics of the rich countries, this necessarily implies that damages are valued less in the poor countries (in the same way that the poor communities would be willing to host a hazardous waste facility for less compensation). From the consequential or substantive rationality inherent in the individualistic perspective, this demonstrates an important concern with trade-offs. It means that other investments would be relatively more attractive to the publics of the poor countries – for example, investments in schools and public health. Schelling's argument for reassessing how we help the world's poor follows from this rationality.

Not surprisingly, when one examines other culturally defined views of fairness, the idea of valuing damages less in the poor countries is objectionable, even immoral. Many groups vehemently object to the WG II report's placing a lower value on a life in poor countries, or what the Global Commons Institute called a calculus that 'values a European as equal to ten China men' (Rayner, 1997). Those advocating egalitarian principles cannot be convinced of the morality of this procedure on the basis of revealed preferences or getting the priorities right. One cannot justifiably set priorities based on an unequal starting point, just as one cannot locate a hazardous waste facility in a poor community. The Pareto rationality is invalidated by the inequities in the world.

The authors of the IPCC chapter grounded their discussion in market utilitarianism, which ultimately distracted from their discussion of values and discredited the analysis of the working group – in much the same way as the World Bank memorandum by Summers discredited the economics of hazardous waste transfers, or that concepts of compensation have discredited voluntary siting schemes. This market–utilitarian view is valid and legitimate, but it is not held by everybody. Explicitly recognizing the different views of fairness suggests a more productive role for the analyst in devising arguments and policies that have wider appeal and legitimacy.

In some ways, the discussion of obligations and emission rights has proceeded in this direction, although with no explicit recognition of the cultural perspectives or empirical data on the plural views of fairness (for a discussion, see IPCC, Chapter 3, 1995). One group of proposals is based on a combination of population and current emissions, thus combining egalitarian with individualistic notions of status quo and willingness to pay (Grubb et al, 1992; Shue, 1993; Welsh, 1993). Solomon and Ahuja (1991) have proposed a formula for allocating emission rights based on three criteria:

- population pegged to a recent year so as not to give an incentive for population growth;
- GDP with allowances for growth but with incentives for emission reduction; and
- on historical emissions.

Cline (1992) also proposes a three-tiered system but weighted somewhat more to individualistic concerns and developed country interests. He proposes a formula based on population, GNP and current emissions. In different ways, each of these proposals captures the concerns of the three active cultures, but each weights differently the egalitarian concerns and developing country interests.

CONCLUSIONS

Drawing on experience from hazardous waste disposal, this discussion has argued that:

- There is no universally valid set of moral principles that can inspire effective personal, institutional or national commitments to a fair distribution of environmental risk burdens, including the burdens of mitigating and coping with climate change.
- Views on the fair allocation of environmental burdens appear to be related to interests, and also to be culturally relative or tied to worldviews and social solidarities *within countries*.
- These views can be empirically verified.

Identifying criteria for designing agreements on climate change that command international and domestic support would appear to be more complex than for siting hazardous waste facilities. But, as anthropologists point out, the variations in political attitudes and values within countries are at least as great as between countries (Thompson et al, 1990). If views on fairness are associated with different cultural perspectives and worldviews, as this discussion proposes, then the analyst can provide data useful for national representatives in negotiating climate change policies that have improved chances of ratification and implementation. Building on the Austrian hazardous waste survey, empirical evidence on the public's and national policy-makers' views on equitable agreements (as well as their interests) can potentially be helpful in identifying a core set of principles that command support in the politics of the climate-change debate.

The international critique of the IPCC *Second Assessment Report*, which valued lives differently in the North and in the South, might have been avoided had the report dealt more vigorously with the moral pluralism apparent in valuing the costs and benefits of climate abatement policies. This critique, perhaps more than any other issue in the climate debate, confirmed the need for systematic policy research on perceptions of fairness held by the different national and international policy actors. As another case in point, had analysts provided a better understanding of the US politics and discourse on global equity, which emphasized mutual responsibility between the North and South, an agreement more amenable to ratification might have emerged from Kyoto. The analyst might also have contributed to an understanding of the ideological cleft that appeared between Europe and the US.

Fairness is thus an issue of empirical diversity. Only by granting legitimacy to the different principles and moral discourses will it be possible to construct further climate agreements that, by compromising but not abandoning the contending notions of fairness, command wide support. Given the large stakes involved, and the sceptical prognosis for ratification of the Kyoto agreement, there is a strong case for more analytical attention to identify the plural views of fairness and the terrain for agreement.

REFERENCES

Agarwal, A and Narain, S, 1991, *Global Warming in an Unequal World*, Centre for Science and Environment, New Delhi

Bertram G, 1992, Tradable emission permits and the control of greenhouse gases, *Journal of Development Studies* 28:3

Baron, J, Gowda, R and Kunreuther, H, 1993, Attitudes toward managing hazardous waste: What should be cleaned up and who should pay for it?, *Risk Analysis* 13:183–192

Boal, M, 1997, Gore's Greens, *The Village Voice*, Jan 20

Bodansky, D, 1993, The United Nations Framework Convention on Climate Change: A commentary, *Yale Journal of International Law* 18:453–558

Bolin, B, 1998, The Kyoto negotiations on climate change: A science perspective, *Science* 279:330–331

Bullard, R D, 1993, Waste and racism: A stacked deck? *Forum for Applied Research and Public Policy* Spring:29–35

Cline, W R, 1992, *The Economics of Global Warming*, Institute for International Economics, Washington, DC

Douglas, M, 1985, *Risk Acceptability According to the Social Sciences* Russell Sage Foundation, New York

Easterbrook, G, 1997, Hot air treaty, *US News*, Dec 22

Easterling, D, 1992, Fair rules for siting a high level nuclear waste repository, *Journal of Policy Analysis and Management* 11(3):442–475

Elster, J, 1992, *Local Justice: How Institutions Allocate Scarce Goods and Necessary Burdens*, Russell Sage Foundation, New York

Epstein J and Gupta, R, 1990, *Controlling the Greenhouse Effect: Five Global Regimes Compared*, Brookings Institution, Washington, DC

Frey, B S, Oberholzer-Gee, F and Eichenberger, R, 1995, The old lady visits your backyard: A tale of morals and markets, Paper presented at the Workshop on Application of Economics, University of Chicago

Gerrard, M B, 1994, *Whose Backyard, Whose Risk: Fear and Fairness in Toxic and Nuclear Waste Siting*, MIT Press, Cambridge

Grubb, M J, 1990, *Energy Policies and the Greenhouse Effect, Volume 1: Policy Appraisal*, Royal Institute of International Affairs, London

Grubb, M, Sebenius, J, Magalhaes, A and Subak, S, 1992, Sharing the burden, in: I M Mintzer, ed, *Confronting Climate Change: Risks, Implications and Responses*, Cambridge University Press, Cambridge

Grübler, A and Fujii, Y, 1991, Intergenerational and spatial equity issues of carbon accounts, *Energy* **16**:1397–1416

The Guardian, 1992, Lawrence Summers: Why the rich should pollute the poor, Feb 2, p 8

IPCC (Intergovernmental Panel on Climate Change), 1995, *Equity and Social Considerations*, Chapter 3, Working Group III, (lead authors: T Banuri, K Goran-Maler, M Grubb, H K Jacobson, and F Yamin), Draft 8.3

Johnson, A, 1998, The influence of political culture on the formation of pre-regime climate change policies in Sweden, the United States and Japan, *Environmental Values* **7**:223–44

Kemp, R, 1990, Why not in my backyard? A radical interpretation of public opposition to the deep disposal of radioactive waste in the United Kingdom, *Environment and Planning* **22**:1239–1258

Kukathas, C and Pettit, P, 1990, *Rawls: A Theory of Justice and its Critics*, Polity Press in association with Blackwell Publishers, Cambridge

Kunreuther, H, Kleindorfer, P, Knez, P and Yaksick, R, 1987, A compensation mechanism for siting noxious facilities: Theory and experimental design, *Journal of Environmental Economics and Management* **14**(4):371–383

Kunreuther, H, Fitzgerald, K and Aarts, T, 1993, Siting noxious facilities: A test of the facility siting credo, *Risk Analysis* **13**:301–318

Kymlicka, W, 1990, *Contemporary Political Philosophy: An Introduction*, Clarendon Press, Oxford

Laws, D and Susskind, L, 1991, Changing perspectives on the facility siting process, Draft paper, Massachusetts Institute of Technology, Boston

Lidskog, R and Elander, I, 1992, Reinterpreting locational conflicts: Nimby and nuclear waste management in Sweden, *Policy and Politics* **20**:249–264

Linnerooth-Bayer, J and Davy, B, 1994, Hazardous waste cleanup and facility siting in Central Europe: The Austrian case, Report to the Bundesministerium für Wissenschaft und Forschung, IIASA Contract No 93–105, International Institute for Applied Systems Analysis, Laxenburg, Austria

Linnerooth-Bayer, J and Fitzgerald, K, 1994, Conflicting views on fair siting procedures: evidence from Austria, Paper presented at the IIASA Conference on Fairness and Siting, May 23–24, Laxenburg, Austria

Linnerooth-Bayer, J and Fitzgerald, K, 1996, Conflicting views on fair siting processes, *Risk: Health, Safety & Environment* **7**:109–119

Linnerooth-Bayer, J and Löfstedt, R E, 1996, Transboundary Environmental Risk Management: An Overview, Paper presented at the Conference on Transboundary Environmental Risk Management, Warsaw, 6–8 October, 1996

MacLean, D, 1993, Variations on fairness, Paper presented at the IIASA Conference on Risk and Fairness, International Institute for Applied Systems Analysis, June 20–22, Laxenburg, Austria

March, J G and Olsen, J P, 1989, *Rediscovering Institutions: The Organizational Basis of Politics*, Free Press, New York

Mitchell, R C and Carson, R T, 1986, Property rights, protest, and the siting of hazardous waste facilities, *AEA Papers and Proceedings* **76**:285–290

O'Hare, M, Bacow, L, and Sanderson, D, 1983, *Facility Siting and Public Opposition*, Van Nostrand, New York

Parikh, J K, 1995, Joint implementation and North–South cooperation for climate change, *International Environmental Affairs* **7**:22–41

Patt, A, 1997, Economists and Ecologists: Different Frameworks for Analyzing Global Climate Change, IIASA Interim Report, International Institute for Applied Systems Analysis, Laxenburg, Austria

Pearce, D W, Cline, W R, Achanta, A N, Fankhauser, S, Pachauri, R K, Tol, R S J and Vellinga, P, 1996, The social costs of climate change, in: *The Second Assessment Report of Working Group III*, Cambridge University Press, Cambridge

Portney, K, 1985, The potential of the theory of compensation for mitigating public opposition to hazardous waste treatment facility siting: Some evidence from five Massachusetts communities, *Policy Studies Journal* **14**:81–89

Rayner, S and Cantor, R, 1987, How fair is safe enough? The cultural approach to societal technology choice, *Risk Analysis* **7**(1):3–13

Rayner, S, 1994, A conceptual map of human values for climate change decision making, Paper presented at the Workshop on Equity and Social Considerations, IPCC Working Group III, Nairobi, Kenya

Rayner, S, 1997, *Integrated climate change assessments: fairness and equity issues, an overview*, Battelle, Pacific Northwest National Laboratory, Washington DC

Rayner, S and Malone, E, 1998, *Human Choice and Climate Change: Ten Suggestions for Policy Makers*, Battelle Memorial Institute, Washington, DC

Rawls, J, 1971, *A Theory of Justice*, Harvard University Press, Cambridge, MA

Rescher, N, 1966, *Distributive Justice: A Constructive Critique of the Utilitarian Theory of Distribution*, Bobbs-Merrill, Indianapolis, IN

Schelling, T C, 1994, Intergenerational discounting, *Energy Policy* **23**:23–32

Schelling, T C, 1998, Costs and benefits of greenhouse reduction, *Issues in Science and Technology*, forthcoming

Schwarz, M and Thompson, M, 1990, *Divided We Stand*, University of Pennsylvania Press, Philadelphia

Shue, H, 1993, Subsistence emissions and luxury emissions, *Law and Policy* **15**:1

Solomon, B D and Ahuja, D R, 1991, International reductions of greenhouse-gas emissions: An equitable and efficient approach, *Global Environmental Change* **1**:5

Thompson, M, Ellis, R and Wildavsky, A, 1990, *Cultural Theory*, Westview Press, Boulder, CO

Thompson, M, 1994, Unsiteability: What should it tell us?, *Risk: Health Safety & Environment* **7**:169–179

Tumulty, K, 1997, A treaty meets a sour congress, *Time*, **150**:26

Van Well, L, 1997, Process and substance: theories of justice and the global climate change debate, draft paper, The Swedish Institute of International Affairs, Stockholm

Victor, D G, 1999, The regulation of greenhouse gases – Does fairness matter?, (Chapter 12 of this book)

Welsh, H, 1993, A CO_2 agreement proposal with flexible quotas, *Energy Policy* **22**(July)

Wilson, J, 1992, The moral sense, *American Political Science Review* **87**(1):1–11

World Energy Council, 1997, *COP-3 to the UN Framework Convention on Climate Change and the Kyoto Protocol*, World Energy Council, London

Yellen, J, 1998, Report given to the U S House Commerce Subcommittee, Reuters

Young, P, 1994, *Equity in Theory and Practice*, Princeton University Press, Princeton, New Jersey

Zemanek, K, 1992, State responsibility and liability, environmental protection and international law, in: W Lang, H Neuhold, and K Zemanek, eds, Graham and Trotman, UK

4 EMPIRICAL AND ETHICAL ARGUMENTS IN CLIMATE CHANGE IMPACT VALUATION AND AGGREGATION

Richard S J Tol, Samuel Fankhauser and David W Pearce

INTRODUCTION

A reasonable understanding of the likely impacts of climate change is crucial for making an informed decision about the best response strategy to the enhanced greenhouse effect. For this reason, the *Second Assessment Report* (SAR) of the Intergovernmental Panel on Climate Change (IPCC) has paid particular attention to the study of impacts. Working Group II has extensively reviewed the literature on the physical impacts of climate change (Watson et al, 1996).

However, it may be useful to aggregate the vast number of regional and sectoral estimates to a more easily tractable set of numbers. Money is the most commonly chosen numeraire for such aggregation. Monetization has the advantage of allowing comparison to the cost of response measures, and to other issues (for example, acidification, health care). The social costs chapter (Chapter 6) of the IPCC Working Group III SAR has reviewed the available monetary assessments of climate change damages (Pearce et al, 1996).

The discussion on the social cost chapter (Bruce, 1995; Masood, 1995; Masood and Ochert, 1995; Meyer, 1995) highlighted many of the difficulties of monetary valuation, particularly its ethical underpinnings. Some of the critique, however, may have been inspired by the perceived underestimation of the impact.[1]

This chapter draws on Fankhauser and Tol (1996, 1997), Fankhauser et al (1997, 1998) and Tol and Fankhauser (1998). It is structured as follows. The second section briefly reviews the economic assessment of climate change damages as presented in Pearce et al (1996). The third section addresses the

---

1 Although the annual damage equivalent to 1.5 per cent of world GDP can hardly be considered low. New studies, notably Mendelsohn et al (1997), suggest this figure is too high.

choice between willingness to pay (WTP) and willingness to accept (WTA), the extrapolation of local case study estimates to other regions, and equity weights. The fourth section discusses some of the methods to avoid regionally differentiated per unit damage estimates. Conclusions are drawn in the fifth section. The tension between empirical realities and ethical desirabilities is manifest in each section. In a number of occasions, there is also tension between rash and more careful moral considerations.

CLIMATE CHANGE DAMAGE COSTS

Information on the impacts of global warming is available for several regions and countries. The best studied regions are developed countries, in particular the United States (Smith, 1996) and the United Kingdom (UKCCIRG, 1996). Studies for other regions are scarce and scattered.

Studies usually deal with only a subset of damages and are often restricted to a description of impacts in physical terms. Estimates generally combine, but do not neatly separate the costs of adaptation (such as sea-level rise protection) and the costs of residual damages (such as the inundation of unprotected areas).

By far the best studied impact categories are agricultural impacts (Reilly et al, 1996) and the costs of sea level rise (Bijlsma et al, 1996). Several types of impact have largely been ignored so far because they could not be sufficiently quantified. Other damages were estimated on the back of an envelope.

Attempts at a comprehensive monetary quantification are relatively rare and usually restricted to the US (Cline, 1992; Titus, 1992; Mendelsohn and Neumann, 1998; Nordhaus, 1991). Preliminary estimates of monetary damage in different world regions are provided by Fankhauser (1995), Tol (1995) and Mendelsohn et al (1997). Valuation is based on a mix of WTP, occasionally WTA, and various approximations, including benefit transfers. The Fankhauser and Tol figures, which were at the core of the IPCC assessment (Pearce et al, 1996), are reproduced in Table 4.1. Fankhauser and Tol (1997) have recalculated the initial set of estimates, consistently correcting for purchasing power parity and using the same benefit transfer methodology throughout.[2] These results are reproduced in Table 4.1. The estimates in Table 4.1 reflect the annual impact a doubling of the atmospheric concentration of carbon dioxide would have on the current society.

Figures vary between 0 and 7 per cent of GDP. Table 4.1 highlights the substantial differences between regions. For the former Soviet Union, for example, damage could be as low as 0.4 per cent of GDP, or even negative (climate change is beneficial). Asia and Africa, on the other hand, could face rather high damages, mainly due to the severe life/morbidity impacts. Developing countries generally tend to be more vulnerable (in relative terms) to climate change than developed countries because of the greater importance of agriculture, lower health standards and the stricter financial, institutional and knowledge constraints on adaptation.

---

2 The estimates in Pearce et al (1996) are based on a mixture of market exchange rates and purchasing-power parity exchange rates for the impact, expressed as a percentage of gross domestic product at market exchange rates. See Fankhauser and Tol (1997).

Table 4.1 Annual monetized 2 × CO$_2$ damage in different world regions

	Fankhauser		Tol	
	billion US$	%GDP	billion US$	%GDP
European Union	63.6	1.4		
United States	61.0	1.3		
Other OECD	55.9	1.2		
OECD America			74.5	1.5
OECD Europe			57.4	1.6
OECD Pacific			60.7	3.8
Total OECD	180.5	1.3	192.7	1.9
E. Europe / Former USSR	29.8[a]	0.4[b]	−14.8	−0.4
Centrally Planned Asia	50.7[b]	2.9[b]	−4.0	−0.1
South and South East Asia			92.2	5.3
Africa			46.4	6.9
Latin America			40.3	3.1
Middle East			11.5	5.5
Total Non-OECD	141.6	0.9	172.8	1.7
World	322.0	1.1	364.4	1.8

a Former Soviet Union only
b China only
Source: Fankhauser and Tol, 1997

Pearce et al (1996) stress the preliminary and incomplete character of these estimates. In particular, it should be noted that the above figures are *best guess* estimates. The range does not reflect a confidence interval, but the variation of estimates found in the literature.[3] There is a considerable range of error which has not been quantified. Pearce et al also note that figures on developing countries in particular are largely based on approximation and extrapolation, and are clearly less reliable than those for developed regions. Further, as best guess estimates, the figures neglect the possibility of impact surprises (such as social and political unrest), and of low probability/high impact events (such as a slow down of the thermohaline circulation). To avoid long-term predictions, damage figures measure the impact of 2 × CO$_2$ on a society with today's structure. Vulnerability is likely to change as regions develop and population grows. Despite these shortcomings, available figures give a rough indication of the possible order of magnitude of climate change impacts and the relative vulnerability of various regions.

---

3 Note that the grey literature includes estimates that are both higher (Hohmeyer and Gärtner, 1992; Meyer and Cooper, 1995) and lower (Mendelsohn et al, 1997).

VALUATION ISSUES

Greenhouse damage estimates still have a number of limitations. Several important issues concern valuation. One is the choice between the two concepts of willingness to pay (WTP) and willingness to accept compensation (WTA), which is essentially an issue of property rights. A second issue is the question of benefit transfer, which asks how estimates for one region or one problem area can be extrapolated to another. A third issue concerns the incorporation of equity issues into comparison and aggregation of estimates.

Willingness to pay versus willingness to accept

It is an empirical fact that economic values derived under a WTP framework tend to differ from estimates that measure the same damage using WTA. Bateman and Turner (1992), for example, report ratios of WTA over WTP ranging from 1.6 up to 6.5. Climate change damages, too, can be expected to be sensitive to the choice of valuation concept.

Various reasons for this discrepancy have been advanced. Firstly, the valuation experiments showing the discrepancy may have failed to replicate near market contexts. When respondents are asked to repeat bids, for example, WTP and WTA eventually converge (Coursey et al, 1987). Secondly, respondents may be rejecting the implied property right. WTA implies that the sufferer does not 'own' the environment, or that the sufferer is bereft of an environmental property. If the respondent feels that is incorrect or immoral, large WTA values may result, including high 'protest bids'. Thirdly, prospect theory suggests that respondents are anchored to a reference point – generally their prevailing bundle of goods and assets. Taking some of these goods away is then treated very differently from adding to the set of goods. This is reinforced if respondents see the good in question as part of their 'identity' (Kahneman and Tversky, 1979). Fourthly, the discrepancy between WTP and WTA is least where the substitution possibilities for goods are highest. Unique assets, often the context of contingent valuation studies, will tend to have high WTP/WTA discrepancies (Hahnemann, 1991).

The choice between WTP and WTA constitutes an implicit statement about prevailing property rights, and this is sometimes used as a guideline for the choice of concept. By using WTP – asking people how much they would pay to avoid adverse impacts – a changing climate is implicitly chosen as the reference scenario. People do not have a 'right' to the climate currently observed, but have to pay to obtain it. Conversely, by using WTA, the assumption is that people are entitled to the current climate. They have to be compensated for any damage arising from alterations to it.

However, the appropriate allocation of property rights (and thus the choice between WTP and WTA) in the case of climate change is unclear. On the one hand, the right of future generations to a functioning environment seems hard to question. This would point towards WTA: future generations are to be compensated for a climate-change induced deterioration in living standards. On the other hand, an equally strong case can be made for the right of developing countries to increase their standard of living, which would imply at least a certain degree of baseline warming, and hence the use of WTP. Similarly, the benefits of past growth and current economic strength in the countries of the OECD should not be fully discounted against global warming.

In practice, WTP and WTA are often mixed up. This is particularly the case for climate change damages, where the limited number of original studies makes it necessary to use whatever information is available, often resorting to benefit transfer. Most estimates in the literature were derived with the benefits of emission reductions in mind, and consequently took business-as-usual climate change as the starting point, asking people about their WTP to obtain a deviation. WTA has also been used, however, as have been a number of second-best measures that were used as approximations in the absence of primary studies.

The distinction between WTP and WTA is thus blurred for current impact estimates. Nevertheless, the issue is important conceptually, and it is likely to become increasingly relevant as impact studies are refined.

Benefit transfer and scaling by income

The climate-change damage cost literature relies heavily on benefit transfer. This is a short cut, in which WTP/WTA results obtained in one study are transferred to a new problem. For the assessment of climate-change induced mortality risk, for example, per unit values were 'transferred' from value of a statistical life (VOSL) studies in various developed countries to all countries.

Benefit transfer is not without problems. Estimates are often site or problem specific and hence difficult to transfer. The cause of the mortality risk may differ. If WTP/WTA depends on the type of hazard, the transferred per unit value would be biased. It is conceivable that the WTP/WTA for mortality risks varies with age. Risks of climate change are more imposed (as opposed to voluntary) than most other risks and less attributable. For other climate change related risks (for instance, malaria, or extreme events), the WTP/WTA may be different again. In some cases, climate change will slightly alter current risks; in other cases, climate change will introduce entirely new risks.

To take such effects into account, the values from the underlying study should ideally be corrected for differences in site and socioeconomic conditions. Even so, the accuracy of benefit transfer remains open to question (Bergland et al 1995; Alberini et al, 1995).

In most climate change damage studies, the only adjustment made concerned income, which is one of the main explanatory variables for both WTP and WTA. A standard assumption is that WTP/WTA is an increasing function of income.[4] A rich person would normally be willing (and able) to make a higher payment, in absolute terms, than a poor person. By the same token, a compensation of, say, US$1000 will appear less attractive to a rich person than to a poor individual. The damage studies reviewed in Pearce et al (1996) therefore scale per unit values according to income – they use lower values in low income countries.

---

4 Although the theoretical work of Flores and Carson (1997) does not exclude the possibility of a negative correlation.

Although there is clear empirical evidence for an income effect, little is known about its magnitude, and scaling is correspondingly controversial. Most of the studies surveyed in Pearce et al (1996) assume an income elasticity of WTP/WTA of one. That is, WTP/WTA as proportions of income are identical across individuals. If a rich person is willing to pay, say, 5 per cent of his income for an environmental good, a poor person would equally be willing to spend 5 per cent of his.

Recent results suggest an income elasticity of less than one (Flores and Carson, 1997; Kristrom and Riera, 1996). A lower income elasticity would imply that damages in developing countries were underestimated initially. (The estimates for developed countries are not affected, since these are the subject of the original study.) The evidence is not yet conclusive, though. The few available studies that directly estimate the VOSL in developing countries all came up with substantially lower values than would be obtained through benefit transfer with an income elasticity of unity. Thus, Parikh et al (1994) found VOSLs in Bombay of US$15,000 to $25,000. Da Motta et al (1993) find a VOSL of US$15,000 using the human capital approach in Brazil. For comparison, the lowest VOSL assumed in the IPCC social cost chapter is US$150,000. Note, however, that, since based on the human capital approach, it is likely that the Indian and Brazilian studies have underestimated the true VOSL.

Better information is clearly needed in this area. At the same time, it should be recalled that the question of the income elasticity of WTP/WTA has arisen only due to the absence of original damage research which necessitated the use of benefit transfer. Although benefit transfer is likely to continue to be important, primary studies directly concerned with the valuation of climate change damages are therefore at least as important.

Equity weighting

WTP/WTA estimates are a mixture of descriptive and prescriptive concepts. The value of goods is set according to people's own appreciation (description), and people's preferences should count (prescription). At the same time, the socioeconomic situation from which people make their assessment is taken as a given. This can lead to problems if the currently observed situation (say, the distribution of income) is considered to be unfair. WTP/WTA estimates will reflect this unfairness. Not least for this reason, some authors have called for the use of uniform per unit damage values (see the fourth section of this chapter). The solution offered by welfare economics is not to use uniform per unit values, but to weight individual estimates by a corrective factor that adjusts values for inequalities in the income distribution.

Fankhauser et al (1997) calculate such 'equity weights' and the corresponding damage figures for a variety of possible welfare functions.[5] Worldwide damage can be expressed as:

$$D^{world} = \left(\frac{W_1 \cdot u_Y^1}{W_M}\right) D^1 + \ldots + \left(\frac{W_n \cdot u_Y^n}{W_M}\right) D^n \qquad (1)$$

---

[5] Azar and Sterner (1996) also use equity weights. Their impact figures are assumptions rather than estimates.

where the terms in brackets denote the equity weights. Equity weights consist of three parts. Firstly, u_Y is a measure how much regional welfare changes as a consequence of damage D. Secondly, W_i is a measure of how much global welfare changes as a consequence of a change in regional welfare. Thirdly, W_M is a scaling factor which ensures that equity weights are in unity if the underlying income distribution or the distribution of damages is considered just.

Three debatable assumptions underlie equation 1. Firstly, meaningful welfare functions do exist. Secondly, economic and environmental goods and services are substitutable, at least within the stress imposed by climate change. Thirdly, climate change impacts are small enough to allow for linearization. These assumptions are needed for a rigorous derivation of equity weights. Future work may look at the implications of these assumptions.

A selection of results is reproduced in Table 4.2. A conventional iso-elastic utility function:

$$u = \frac{a}{(1-e)} \cdot Y^{(1-e)} \tag{2}$$

is used. Different values for parameter e (the income elasticity of marginal utility, or risk aversion) will be used below.

The specification for the welfare function is:

$$W = \frac{\sum_{i=1}^{n} u^i (\cdot)^{(1-\gamma)}}{1-\gamma} \tag{3}$$

where γ is a parameter of inequality aversion. The larger γ is, the larger is the concern about equality. For $\gamma = 0$, equation 4 reduces to a utilitarian welfare function, $\gamma \rightarrow 1$ gives a Bernoulli-Nash function, $\gamma \rightarrow \infty$ represents the maximin (Rawlsian) case, and $\gamma \rightarrow \infty$ is the maximax (Nietzschean) welfare function.

As Table 4.2 makes clear, estimates are highly sensitive to the assumed welfare concept. World damages are generally (but not necessarily) higher than reported in Pearce et al (1996). This exercise also makes clear that the original studies (using equity weights of one) implicitly assumed either a just distribution of welfare or a linear, utilitarian welfare function.

THE PURSUIT OF UNIFORM PER UNIT VALUES

Regional differentiation of per unit values is common practice in valuation, and a direct consequence of the WTP/WTA approach. Regionalized estimates would be closer to the reality of national decision-makers. Nevertheless, some consider the notion of regionally differentiated values as unethical, particularly for mortality risks.

The valuation of mortality risks (or 'statistical lives') is difficult and controversial. The 'value of life' as such is not valued, rather people's willingness to pay to avoid (or willingness to accept compensation for) an increased risk of premature death. A VOSL expresses the aggregation of individuals' valuations of risk reduction (or increase).

Table 4.2 Aggregate damages corrected for inequality (in billion US$)

	Fankhauser (1995)	Tol (1995)
Uncorrected damages[a]	322.0	364.4
Utilitarian Welfare Function		
$e = 0.0$[b]	322.0	364.4
$e = 0.5$	315.6	411.4
$e = 1.0$	405.2	614.3
$e = 1.5$	621.9	1057.6
$e = 2.0$	1041.7	1930.0
Bernoulli-Nash Welfare Function[c]	405.2	614.3
Maximin Welfare Function		
$e = 0.0$	50.7	46.4
$e = 0.5$	95.8	89.4
$e = 1.0$	181.0	172.2
$e = 1.5$	342.7	331.8
$e = 2.0$	646.5	639.3

a as in Table 4.1
b e denotes the income elasticity of marginal utility
c Bernoulli-Nash weights are independent of e, and correspond to the case $e = 1$ of the utilitarian welfare function.
Source: Fankhauser et al, 1997

Above, VOSLs are regionally differentiated. Several alternatives have been put forward, generally aiming at a uniform valuation of mortality risks for all individuals.

Averaging

Evidently, WTP/WTA estimates do not only differ between nations, but also between different socioeconomic groups within one country. Sometimes, nationally averaged values are used, for ease of calculation, and to assure the acceptability of the assessment to all stakeholders. If necessary, compensatory policies to restore inequities resulting from such valuation may be installed.

For the same reason of acceptability, the use of globally averaged values has been advocated for the assessment of climate-change induced (mortality) risks. In doing so, the result is that the revised figures for global mortality damage remain more or less unchanged from Pearce et al (1996). The reason is this. Both Fankhauser (1995) and Tol (1995) assume mortality to be directly proportional to population. Averaging, therefore, has no effect on total damage. The linearity assumption is now increasingly called into question (Kalkstein and Tan, 1995; Martens et al, 1995).

If the analysis relies heavily on benefit transfer (as it does), one may prefer not to use the average of the inferred VOSLs, but the VOSL associated with the average per capita income. The two values differ if VOSL is not proportional to income. Both Fankhauser (1995) and Tol (1995) assume VOSL to be close to proportional per capita income.

The use of average values seems to be a pragmatic way forward in general. However, it may lead to inconsistencies in the way locally and globally caused damages are measured.[6] Evidently, environmental problems of local origin (say, related to air pollution) would continue to be assessed at local per unit values. The result may then be large discrepancies in valuation, depending on the origin of the risk. The incremental mortality risk from climate-change induced malaria, for example, would be valued differently than, say, the same increase in risk caused by deteriorating medical standards. Similarly, traffic-related mortality risks would be valued differently than heat-stress related risks. This may lead to an unjustified reallocation of expenditures on safety.

Maximum value

The use of common averages for VOSL in climate valuation studies would be more or less consistent with national approaches to the use of VOSL. What would not be consistent, however, is the use of the highest average number observed in any country in the world. This approach has been recommended by Meyer and Cooper (1995) and Hohmeyer and Gärtner (1992) on the grounds that no other value would be acceptable to the country with the highest VOSL.

Their argument is untenable for the same reasons why average valuation is untenable, albeit much stronger. Using a maximum value for a statistical life would lead to an overemphasis of climate change relative to other mortality risks for all countries but the one with the highest VOSL.

Historical responsibility and polluter pays

The uniform use of per unit values at OECD level has also been advocated by Ekins (1995) and Hohmeyer (1996). In an aberration of the polluter pays principle, they argue that, because OECD countries are predominantly responsible for the accumulation of greenhouse gases in the atmosphere, it is justifiable to value all damages at the per unit values observed in the OECD countries.

Perhaps the main problem with this approach is the unfortunate combination of valuation (essentially an empirical matter) with the political question of equity and compensation. The polluter pays principle provides a rule about the *direction* of compensation payments (from the polluter to the victim). It does not determine the *magnitude* of the payment, although payments should reflect damages incurred. The principle says who pays for the incurred damage, not *how much* is paid.

A second problem is that this approach reverses common practice in litigation cases, where compensatory payments are determined according to the damage suffered by the victim. Instead, damage is determined here by the welfare loss the offender would have suffered, had he or she been the victim.

---

6 Note that consistency is not a prime aim of real world policy. In our opinion, it should be an aim of policy advice. Accountability implies consistency. Accountability may be an aim of policy.

A third problem is that estimates now crucially depend upon who is perceived to be responsible for which part of the inflicted damage. Assigning responsibility is a complicated matter. At first sight the facts are clear: the countries of the OECD have contributed most to the accumulation of greenhouse gases in the atmosphere. What is more, they were first warned by Arrhenius in 1896. Nevertheless, making OECD countries solely responsible for the entire greenhouse effect may be too easy. Collective responsibility is not the same as individual responsibility. Until the mid 1980s, countries could hardly be held accountable for ignoring an old paper. Since then, each country is severely restricted by the need for a broad international coalition to successfully abate climate change. In addition, careful consideration ought to be given to the question whether combating climate change does not create other problems,[7] or would have created had greenhouse gas emissions been abated from the early 1900s onward. Thus, the perception of responsibility strongly depends upon one's ethical position as well as upon the question whether one is led by what is or by what ought to be.

For the purpose of illustration and sensitivity analysis, Table 4.3 shows the 'damages' that would result from the Ekins and Hohmeyer/Gärtner approach. It also distinguishes between assigning responsibility to OECD and Annex I countries. The table also distinguishes between various degrees of responsibility. The OECD or Annex I is either held fully responsible for all damages; or the OECD or Annex I is only responsible for its own damages, plus a fraction of damage in poor countries equal to its historic contribution to greenhouse gas emissions. The first column of Table 4.3 reproduces Fankhauser's estimates. In the second column, the rich region is held responsible for all impacts; hence, all damage in the poor region is valued as the level of the rich. In the third column the rich region is only partially accountable for damages in the poor region. World damage estimates increase by about a factor of 2.5 to 4 as a consequence of these assumptions. Damages in developing countries increase by an even larger factor. It is important to note, though, that these figures no longer reflect the actual welfare loss people will face.

The last three columns of Table 4.3 adopt the polluter pays principle: responsible regions compensate others for the damage caused. Costs consist of self-inflicted damages, as well as compensation paid, minus compensation received. Damage is again valued at regional per unit values, so that world damage remains unaltered. In column 4, the rich are held accountable for all the damage suffered anywhere in the world, and hence meet all damage costs. Net costs to the poor drop to zero. In column 5, the rich are held partially accountable (and hence only partially compensate) for damage in poor regions. The sixth column analyses reciprocal responsibility. That is, the rich compensate the poor, and the poor the rich, according to their respective contributions to the enhanced greenhouse effect. This last scenario provides results closest to the original figures.

---

[7] This may be the rationale for lower commitments for the non-Annex (II) countries under the Climate Convention, even though the consequences of their greenhouse gas emissions are today better and more widely known than ever.

Table 4.3 Fankhauser's estimates for different positions on valuation and accountability

Region	Original[a]	Valuation[b] all[d]	64/84[e]	Accountability[c] all[d]	64/84[e]	64/84[f]
OECD	1.36	1.36	1.36	2.42	2.04	1.55
non-OECD	0.86	6.86	4.39	0.00	0.31	0.70
World	1.08	4.41	3.04	1.08	1.08	1.08
Annex-I	0.99	0.99	0.99	1.52	1.44	1.28
non-Annex	1.29	7.77	6.52	0.00	0.21	0.60
World	**1.08**	**2.96**	**2.60**	**1.08**	**1.08**	**1.08**

a As in Table 4.1.
b Values in the poor region (non-OECD and non-Annex, respectively) are adjusted to the values in the rich region (a factor of 8 and 6, respectively).
c The rich region (OECD and Annex-I, respectively) compensates the poor region for damage suffered from climate change.
d Damages adjusted or compensated for all impacts in poor countries.
e Damages adjusted or compensated for 64 per cent (non-OECD) or 84 per cent (non-Annex I) in poor countries. The OECD and Annex-I contributed an estimated 64 per cent and 84 per cent, respectively, to accumulated carbon dioxide emissions from industrial sources (Grübler and Nakicenovic, 1991).
f Rich and poor compensate one another for damage suffered from climate change according to their respective accountability as in footnote e.
Source: Fankhauser et al, 1998

Inverse equity weighing

Equal values can be analysed in the framework of equity weighing. Equal values then implicitly prescribe a welfare function. The requirement is that the equity weight ratios are exactly the opposite of the value ratios. Using the same set up as above, but restricting ourselves to two income groups, this requirement can be written as:

$$e\gamma - \gamma - e = \Omega \qquad (4)$$

with $\Omega = [ln(V^p)-ln(V^r)]/[ln(Y^r)-ln(Y^p)]$. For any given value for Ω and e (which are empirical facts), the presumption that climate change impacts are to be valued equally implies a certain value for inequality aversion γ.

Table 4.4 presents γ as a function of e and V^r/V^p. We assume values for e between 0.5 and 1.5. The ratio V^r/V^p is more difficult to determine since empirical evidence is scarce. In Table 4.4, values range from 1.4 to 10. An income elasticity of 1 would imply a ratio of 4.

As Table 4.4 shows, the postulate of uniform per unit values is compatible with many sets of 'reasonable' parameter assumptions, but by no means with all of them. For several parameter specifications common values imply degrees of inequality aversion in the utilitarian ($\gamma = 0$) or Bernoulli-Nash range ($\gamma = 1$). In the case of a unitary income elasticity of WTP, for example, uniform per unit values imply a Bernoulli-Nash welfare function. Other parameter sets imply higher degrees of inequality aversion, and in the case of a logarithmic utility

Table 4.4 Implied inequality aversion (γ) as a function of the income elasticity of marginal utility (e) and empirical value ratio (V^r/V^p)

V^r/V^p	1.36	2	4[a]	8	10
γ, for					
$e = 0.5$	−0.56	−0.01	1.00	1.98	2.31
$e = 1.0$	±∞[b]	±∞[b]	1.00	±∞[c]	±∞[c]
$e = 1.5$	2.56	2.01	1.00	0.02	−0.31

a corresponds to the case $V^r/V^p = Y^r/Y^p$.
b $\gamma \uparrow \infty$ for $e \downarrow 1$, and $\gamma \downarrow -\infty$ for $e \uparrow 1$.
c $\gamma \downarrow -\infty$ for $e \uparrow 1$, and $\gamma \uparrow \infty$ for $e \downarrow 1$.
Source: Fankhauser et al, 1997

function ($e = 1$), common values are in the limit only compatible with a maximin welfare function ($\gamma = \infty$).

There are also cases in which common per unit values would imply negative values for g, that is, 'inequality attraction', which could in the limit go to a maximax welfare concept ($\gamma = -\infty$). With certain parameter combinations, it can happen that weighed per-unit damages estimates for the poor region are higher than those for the rich region. The restriction of equal values then favours the rich. Clearly, this would be an indefensible welfare concept, and it would therefore be hard to make a case for common per unit values should these particular parameter values prevail.

CONCLUSIONS

Estimating the social costs of climate change has been a controversial issue. This chapter discusses the key sensitivities of the usual practice in damage estimation as reflected in the IPCC SAR.

Economic valuation of climate change impacts is difficult. Available figures are still incomplete and have to be interpreted with care. Nevertheless, monetary estimates of climate impacts can provide useful information to decision-makers, whether they are used in a cost-benefit context or not.

Estimates need to be based on a sound theoretical framework and make use of the best estimation techniques available at the time. Ad hoc methods, such as valuation according to historical responsibility or valuation at the highest observed level, do not meet this requirement. Although not without problems, the WTP/WTA method is the best technique currently available. The pursuit of equal values for all, which has an intuitive appeal of equity, may have implications that are highly inequitable. The solution advanced here – equity weights – is one that is based on neoclassical welfare economics. Alternative solutions should be explored.

Damage figures obviously have to be of sufficient quality, and there is a clear need for improved estimates. Studies that value the impacts of climate change directly are particularly required, rather than using benefit transfer. Valuation studies are especially needed for non-market impacts and for developing countries. The more primary studies are available, the less analysts will have to rely on difficult and controversial benefit transfers.

In addition to valuation issues, damage estimates suffer from a number of limitations not related to valuation. Improvements are also essential in that respect. Among the most important research topics are the need to move from equilibrium (2 x CO_2) to transient or dynamic damage analysis, and the need for a better understanding of adaptation.

REFERENCES

Alberini, A, Cropper, M, Fu, T M, Krupnick, A, Liu, J T, Shaw, D and Harrington, W, 1995, *Valuing Health Effects of Air Pollution in Developing Economies: The Case of Taiwan*, mimeo, Resources for the Future, Washington, DC

Azar, C and Sterner, T, 1996, Discounting considerations in the context of global warming, *Ecological Economics* 19:169–184

Bateman, I J and Turner, R K, 1992, *Evaluation of the Environment: The contingent valuation method*, GEC WP 92–18, Centre for Social and Economic Research on the Global Environment, Norwich and London

Bergland O, Magnussen, K and Navrud, S, 1995, Benefit transfer: testing for accuracy and reliability, Discussion Paper D–03/1995, Department of Economics and Social Sciences, Agricultural University of Norway, As

Bijlsma, L, Ehler, C N, Klein, R J T, Kulshrestha, S M, McLean, R F, Mimura, N, Nicholls, R J, Nurse, L A, Perez Nieto, H, Stakhiv, E Z, Turner, R K and Warrick, R A, 1996, Coastal zones and small islands, in: R T Watson, M C Zinyowera and R H Moss, eds, *Climate Change 1995: Impacts, Adaptations and Mitigation of Climate Change: Scientific-Technical Analyses – Contribution of Working Group II to the Second Assessment Report of the Intergovernmental Panel on Climate Change*, Cambridge University Press, Cambridge

Bruce, J P, 1995, Impact of climate change, *Nature* 377:472

Cline, W R, 1992, *The Economics of Global Warming*, Institute for International Economics, Washington, DC

Coursey, D L, Hovis, J J and Schulze, W D, 1987, The disparity between willingness to accept and willingness to pay measures of value, *Quarterly Journal of Economics* 102:679–90

Da Motta, R, Mendes, A, Mendes, F and Young, C, 1993, Environmental damages and services due to household water use, Discussion Paper 258, Instituto de Pesquisa Economica Aplicada (IPEA), Rio de Janeiro

Ekins, P, 1995, Rethinking the costs related to global warming: A survey of the issues, *Environmental and Resource Economics* 5:1–47

Fankhauser, S, 1995, *Valuing Climate Change. The Economics of the Greenhouse*, Earthscan, London

Fankhauser, S and Tol, R S J, 1996, Recent advancements in the economic assessment of climate change costs, *Energy Policy*, 24(7):665–673

Fankhauser, S and Tol, R S J, 1997, The social costs of climate change: the IPCC second assessment report and beyond, *Mitigation and Adaptation Strategies for Global Change* 1:385–403

Fankhauser, S, Tol, R S J and Pearce, D W, 1997, The aggregation of climate change damages: a welfare theoretic approach, *Environmental and Resource Economics* 10:249–266

Fankhauser, S, Tol, R S J and Pearce, D W, 1998, Extensions and alternatives to climate change impact valuation: On the critique on IPCC WG3's impact estimates, *Environment and Development Economics* 3:59–81

Flores, N E and Carson, R T, 1997, The relationship between the income elasticities of demand and willingness to pay, *Journal of Environmental Economics and Management* 33:287–295

Grübler, A and Nakicenovic, N, 1991, *International Burden-Sharing in Greenhouse Gas Reduction*, Environmental Policy Division, World Bank, Washington, DC

Hahnemann, W M, 1991,Willingness to pay and willingness to accept: how much can they differ?, *American Economic Review* **81**(3):635–647

Hohmeyer, O, 1996, Social costs of climate change – strong sustainability and social costs, in: O Hohmeyer, R L Ottinger and K Rennings, eds, *Social Costs and Sustainability – Valuation and Implementation in the Energy and Transport Sector*, Springer, Berlin

Hohmeyer, O and Gärtner, M, 1992, *The Costs of Climate Change*, Report to the Commission of the European Communities, Fraunhofer Institut für Systemtechnik und Innovationsforschung, Karlsruhe, Germany

Kahneman, D and Tversky, A, 1979, Prospect theory: an analysis of decisions under risk, *Econometrica* **47**:263–91

Kalkstein, L S and Tan, G, 1995, Human health, in: K M Strzepek, and J B Smith, eds, *As Climate Changes – International Impacts and Implications*, Cambridge University Press, Cambridge

Kriström, B and Riera, P, 1996, Is the income elasticity of environmental improvements less than one?, *Environmental and Resource Economics* **7**:45–55

Martens, W J M, Jetten, T H, Rotmans, J and Niessen, L W, 1995, Climate change and vector-borne diseases – a global modelling perspective, *Global Environmental Change* **5**(3):195–209

Masood, E, 1995, Developing countries dispute use of figures on climate change impact, *Nature* **376**:374

Masood, E and Ochert, A, 1995, UN climate change report turns up the heat, *Nature* **378**:119

Mendelsohn, R and Neuman, J, *The Impact of Climate Change on the US Economy*, Cambridge University Press, Cambridge, forthcoming

Mendelsohn, R, Morrison, W, Schlesinger, M E and Andronova, N G, 1997, A Global Impact Model for Climate Change, draft

Meyer, A, 1995, Economics of Climate Change, *Nature* **378**:433

Meyer, A and Cooper, T, 1995, A recalculation of the social costs of climate change, Working Paper, *The Ecologist*

Nordhaus, W D, 1991, To slow or not to slow: The economics of the greenhouse effect, *Economic Journal* **101**(407):920–937

Parikh, K, Parikh, J, Muralidharan, T and Hadker, N, 1994, *Valuing Air Pollution in Bombay*, mimeo, Indira Gandhi Institute of Development, Bombay

Pearce, D W, 1995, Valuing climate change, *Chemistry and Industry*, 18 December: 1024

Pearce, D W, Cline, W R, Achanta, A N, Fankhauser, S, Pachauri, R K, Tol, R S J and Vellinga, P, 1996, The social costs of climate change: greenhouse damage and the benefits of control, in: J P Bruce, H Lee and E F Haites, eds, *Climate Change 1995: Economic and Social Dimensions – Contribution of Working Group III to the Second Assessment Report of the Intergovernmental Panel on Climate Change*, Cambridge University Press, Cambridge

Reilly, J, Baethgen, W, Chege, F E, van de Geijn, S C, Lin, E, Iglesias, A, Kenny, G, Patterson, D, Rogasik, J, Roetter, R, Rosenzweig, C, Sombroek, W and Westbrook, J, 1996, Agriculture in a changing climate: impacts and adaptation, in: R T Watson, M C Zinyowera and R H Moss, eds, *Climate Change 1995: Impacts, Adaptations and Mitigation of Climate Change: Scientific-Technical Analyses – Contribution of Working Group II to the Second Assessment Report of the Intergovernmental Panel on Climate Change*, Cambridge University Press, Cambridge

Smith, J B, 1996, Standardized estimates of climate change damages for the United States, *Climatic Change* **32**(3):313–326

Titus, J G, 1992, The cost of climate change to the United States, in: S K Majumdar, L S Kalkstein, B Yarnal, E W Miller and L M Rosenfeld, eds, *Global Climate Change: Implications Challenges and Mitigation Measures*, Pennsylvania Academy of Science, Pennsylvania

Tol, R S J, 1995, The damage costs of climate change – toward more comprehensive calculations, *Environmental and Resource Economics* **5**:353–374

Tol, R S J, 1997, The social cost controversy: A personal appraisal, in: A Sors, A Liberatore, S Funtowicz, J-C Hourcade and J L Fellous, eds, *Proceedings of the International Symposium Prospects for Integrated Environmental Assessment: Lessons Learnt from the Case of Climate Change*, pp 35–42, Brussels: European Commission DGXII

Tol, R S J and Fankhauser, S, 1998, On the representation of impact in integrated assessment models of climate change, *Environmental Modelling and Assessment* 3:63–74

UKCCIRG (United Kingdom Climate Change Impacts Review Group), 1996, *Review of the Potential Effects of Climate Change in the United Kingdom*, HMSO, London

Watson, R T, Zinyowera, M C and Moss, R H, eds, 1996, *Climate Change 1995: Impacts, Adaptation, and Mitigation of Climate Change – Scientific-Technical Analysis – Contribution of Working Group II to the Second Assessment Report of the Intergovernmental Panel on Climate Change*, Cambridge University Press, Cambridge

5 APPLYING FAIRNESS CRITERIA TO THE ALLOCATION OF CLIMATE PROTECTION BURDENS: AN ECONOMIC PERSPECTIVE

Carsten Helm[1]

INTRODUCTION

Article 3 of the United Nations Framework Convention on Climate Change (UNFCCC) states that: 'Parties should protect the climate system for the benefit of present and future generations of humankind, on the basis of *equity* and in accordance with their common but differentiated responsibilities and respective capabilities' (emphasis added). However, the UNFCCC is quiet on the criteria according to which the equitability of different policy proposals on the negotiation table should be judged. Therefore, the demand for equity has provided little guidance until now.

This is not surprising because arguments based on equity considerations are often treated with great suspicion. Two popular reservations, which are also common among economists, go as follows: 'equity is merely a word that hypocritical people use to cloak self-interest'; and 'it is so hopelessly subjective that it cannot be analysed scientifically' (Young, 1994:xi). In many respects, negotiations about climate change seem to offer a perfect confirmation of those reservations because nearly every actor – no matter whether it is a low-lying island state or an oil exporter – has defended its policy proposal as the truly equitable one.

This chapter intends to show that there have been some important research efforts by economists towards a consistent analysis of equity and that the developed apparatus can be applied profitably to fairness concerns in climate change.[2] More specifically, it will introduce a small number of general

---
[1] The author would like to thank Frank Biermann and three anonymous referees for helpful comments.
[2] Several monographs have been published recently with the intention to survey and improve our understanding of equity with methods that are commonly used in economics; see Young (1994), Moulin (1995), Brams and Taylor (1996), Roemer (1996) and Kolm (1997). Yet, most of these contributions focus on equity issues on the national level or among individual agents, and only few efforts have been undertaken to apply economic theories of justice to the international level, or even more specifically to problems related to climate change.

equity criteria and explore their consequences for the allocation of emission reductions and associated costs in a climate protection regime (fifth and sixth sections). Previously, it will discuss some of the approaches towards equity which have received particular attention in the report of the International Panel on Climate Change (IPCC, 1996) (second section), will describe some common proposals for a just distribution of emission reduction burdens (third section), and will discuss some general prerequisites for the approach chosen in this chapter (fourth section).

DIFFERENT APPROACHES TOWARDS EQUITY

Climate change raises a number of different equity issues, of which the most important ones are classified in the IPCC's chapter on equity and social considerations as follows (IPCC, 1996:85):

- international equity in coping with the impacts of climate change and associated risks;
- international equity in efforts to limit climate change;
- equity and social considerations within countries;
- equity in international processes; and
- equity among generations.

Certainly, each of these points is highly relevant and deserves a detailed analysis. This, however, is far beyond the scope of this contribution, in particular because one cannot apply the same methodology to the analysis of all equity issues stated above. Therefore, this chapter largely confines itself to the question of 'international equity in efforts to limit climate change' – that is, the just distribution of emission rights and abatement cost, which is one of the most important and at the same time one of the most controversial equity issues in current negotiations on climate protection strategies (IPCC, 1996:103). This has also become clear in the disputes about the involvement of developing countries in a climate protection regime, which gained in intensity during the negotiations of the Kyoto Protocol.

The IPCC report distinguishes five broad traditions to approaching this question: parity, proportionality, priority, classical utilitarianism, and Rawlsian distributive justice (IPCC, 1996:86; see also Young, 1994:8). The last two approaches are not so much concerned with a specific problem, but instead raise the more general question after the 'just social order'.

Classical utilitarianism starts from the assumption of a social welfare function and advocates an allocation that maximizes the sum total of individual agents' utility – Bentham's principle of achieving the greatest good for the greatest number:

$$W_G = U_1 + U_2 + \ldots + U_n, \tag{1}$$

where – applied to the international level – W_G is global welfare and U_i the utility of the population in country i. A common objection against such an additive social welfare function is that it neglects distributionary aspects in so far as extremely different *utility* allocations would be represented by the same welfare index W_G, as long as the sum of the utility levels remains constant. However, this does not imply indifference towards the *wealth* distribution, because the marginal utility of wealth is usually assumed to be higher in poorer countries. Neglecting the production side of the economy for the moment, global welfare maximization would require marginal utility to be equalized in all countries, thereby supporting a rather egalitarian wealth distribution.

A second critique against the social welfare function of classical utilitarianism is that the additive aggregation requires interpersonal, cardinal comparisons of utility – that is, statements such as A's utility from consuming good x is twice as high as B's utility from consuming good y. Today, economists often eschew these kinds of judgements and prefer an intrapersonal, ordinal utility concept, which relies only on statements such as agent A prefers good x to good y.[3]

Rawls, on the other hand, has advocated the distribution principle that the worst off group in the community should be made as well off as possible. This has sometimes been represented by a maximin social welfare function:

$$W_G = min \{U_1, U_2, ..., U_n\} \qquad (2)$$

However, Rawls was not so much looking for a decision that would somehow maximize the social good, but he rather emphasized the process or context in which decisions are made: he was concerned with the establishment of a set of just institutions in which decision-making could take place. In particular, Rawls did not focus on the subjective well-being of the individual agents, but on the means and instruments – the so-called primary goods (Rawls, 1971) – with which they could foster their well-being. However, Rawls himself has argued that his 'difference principle', according to which inequalities can only be justified if they improve the conditions for the worst off individual, cannot be applied to the level of states (Rawls, 1993).

In summary, the additive social welfare function of classical utilitarianism as well as the Rawlsian distributive justice both refer to the just social order, that is, to the just distribution of resources at large. To analyse the equity issues of climate change, we therefore have to take into account the global distribution of wealth so that environmental policies have to be linked to development policies.

An interesting example of how this can be done is provided by Chichilnisky and Heal (1994). They characterize the level of greenhouse gas concentrations in the atmosphere as a 'privately produced public good'. Accordingly, countries' utility levels $U_i(X_i, Q)$ depend on the consumption bundle of private goods X_i and on the quality of the world's atmosphere Q. A trade-off among the two exists in that abatement efforts to increase the quality of the atmosphere are paid for with lower private consumption levels. Furthermore, Chichilnisky and Heal modify the simple utilitarian social welfare function (equation 1 above) by incorporating country-specific welfare weights λ_i, which represent the marginal social welfare of utility increases in the individual countries:

---

[3] But see Harsanyi's (1955) defence of an additive social welfare function, which is based on von Neumann-Morgenstern utility functions.

$$W(X_1,...,X_N,Q) = \sum_{i=1}^{N} \lambda_i U_i(X_i,Q) \qquad (3)$$

Maximization of equation 3 yields that at a Pareto efficient allocation, marginal abatement costs are inversely proportional to the marginal valuation of the private good, at least if lump sum transfers are not feasible. As the marginal utility of consumption is usually regarded to be higher in poorer countries, this would imply that the industrialized countries should shoulder higher marginal abatement costs than developing countries. However, it should be noted that marginal abatement costs are currently much lower in developing countries (IPCC, 1996). Accordingly, the extent to which developing countries have to undertake abatement efforts depends crucially on their marginal utility of consumption relative to that in industrialized countries, and this value is very difficult to determine.

There are a number of reasons in favour of adopting a holistic perspective as is done by approaches which are based on social welfare functions. In particular, it was agreed in the UNFCCC that countries' different ability to pay for climate protection measures should be taken into account, and the marginal social welfare losses of abatement efforts can be used as a good proxy for this. However, there are also good arguments against such an approach. Most importantly, it is likely that disputes about the 'just international order' – the specification of the functional form of the social welfare function, welfare weights λ_i and marginal utilities of income – would make agreement on a 'just climate policy' even more difficult. Nevertheless, any solution, including the one elaborated on in the following sections, will ultimately have to be judged according to its effect on the global distribution of welfare.

The other three equity criteria mentioned by the IPCC are more straightforwardly applied to the distribution of a particular common resource, such as the sink capacity of the atmosphere for greenhouse gases, at least at first sight.[4] *Parity* means that all actors should be treated equally. The criterion of *proportionality* goes back to Aristotle's equity principle and claims that burdens or benefits should be distributed in proportion to the contributions of claimants. Finally, *priority* means that those with the greatest need should be advantaged (Young, 1994:8).

However, these equity principles are relatively meaningless without a precise determination of the perspective or starting point from which they are applied. For example, in the IPCC report it is stated that 'parity is a formula for equal distribution of burdens or benefits' (IPCC, 1996:86). Nevertheless, it is of course a fundamental difference whether the *benefits* of the atmosphere's sink capacity or the *burdens* of emission reductions are shared equally. Similarly, the criteria of proportionality and priority can be applied to such different issues as the extent of climate change impacts and the relevance of oil exports for the national income, leading to completely different results. Even differentiation among the three criteria depends on the perspective. For example, an *equal* per capita allocation is the same as an allocation *proportional* to the countries' population.

---

4 Actually, it would be more appropriate to speak of the sink capacity of the *biosphere*; in this case, *net* emission rights would have to be allocated.

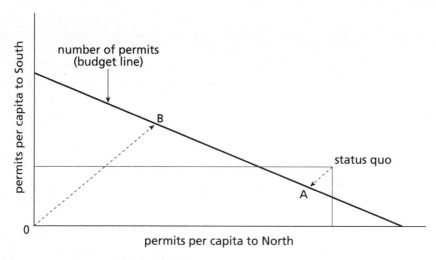

Figure 5.1 Equal distribution of burdens or benefits

Figure 5.1 illustrates the point that not only the choice of the appropriate equity principles is crucial, but also the choice of the perspective from which to apply them. The budget line gives the total number of permits which are allocated between two groups called North and South on the basis of the equity principle of parity. Because total population is higher in the South than in the North, the maximum feasible number of permits per capita is lower in the South. Allocation B says that every country is entitled to emit the same quantity per capita. Allocation A says that every country must reduce its per capita emissions from current levels by the same amount. Even though both allocation procedures are based on the equity principle of parity, the choice of a different reference point leads to completely different results.

COMMON PROPOSALS FOR A 'JUST' DISTRIBUTION OF EMISSION REDUCTION BURDENS

A systematic presentation of the different proposals for a just distribution of climate protection measures is simplified if one separates the question of a just distribution of emission rights from the question of financial transfers (see IPCC, 1996:106–110). The proposals regarding the distribution of emission rights can be classified along the following two extreme positions: equal per capita allocation versus equal percentage reductions. Often, a dynamic allocation scheme has been advocated which starts from a position close to the status quo and moves successively towards an equal per capita allocation (see Cline, 1992:353). Consequently, equity considerations are mixed with those of political feasibility.

Regarding the distribution of abatement cost by monetary transfers, the following approaches can be distinguished:

- Transfers should be paid according to the provisions of the UNFCCC. Of particular relevance is the principle first employed in the ozone regime

that industrialized countries should pay the 'full incremental cost' of developing countries.[5]
- Monetary transfers should result from the distribution of tradeable permits.
- Transfers should be based on the different responsibilities for the environmental problem. This point includes the polluter pays principle and the principle of historical responsibility.
- Transfers should be based on the different ability to pay for climate protection measures.

Supporters of a particular criterion often use one of the above equity principles as justification. Yet, this happens mostly in an ad hoc fashion and is rarely based on a systematic analysis (for a good overview see Grubb et al, 1992; Shue, 1995; IPCC, 1996).

In contrast, this chapter now explores the fruitfulness of an axiomatic approach towards a fair allocation of climate protection measures; the approach is based on five general equity axioms which have received widespread attention in economic theories of justice. However, not only the formulation of the axioms but also the choice of the starting position from which to apply them needs to be justified. This will be done in the following section.

THE CHOICE OF A PERSPECTIVE

Firstly, the fair division problem is treated as one of local justice as opposed to social justice. For the distribution of emission reductions of greenhouse gases and associated reduction costs, this implies an abstraction from the justice of the current global welfare distribution. In contrast to this, some authors have explicitly argued that climate change protection strategies should be designed such that they favour the South (see Simonis, 1996). With the approach followed in this chapter, the author does not deny the legitimacy of such claims but wishes to emphasize that it is important to distinguish whether the justification of the advocated wealth transfers rests on the perceived injustice of the current global welfare distribution or on the characteristics of the climate change problem. On the other hand, the ignorance of superordinate aspects of social justice certainly becomes questionable if the proposed solution for a (local) fair division problem significantly accentuates existing injustices in the social welfare distribution.

Secondly, in concentrating exclusively on justice in efforts to limit pollutive emissions, the author abstracts from other important fairness concerns in climate change, such as justice in coping with the impacts of environmental problems and associated risks. In some cases, emission reduction targets might actually be chosen so that no environmental damages occur. In principle, this has been done in the international regime on stratospheric ozone depletion. Furthermore, the critical loads concept agreed to in the European regime to combat transboundary acidification seeks to reduce emissions to a level below

---

5 See UNFCCC, Art 4 (cf Biermann 1995: 62ff, 106–109).

which significant harmful effects on specified sensitive elements of the environment do not occur (see Biermann, 1995). However, in climate change some damages will not be negligible. The author's intention is not to deny the importance of justice in coping with those impacts, but it is assumed that this and other issues of justice can be analysed separately from the fair division of emission reductions and associated costs.[6]

Thirdly, no intertemporal equity trading is allowed: the allocation has to be just at every time point. This excludes situations where the unfair treatment of agents in one period is compensated for by preferential treatment in other periods. The main reason for this assumption is to simplify the analysis. However, as a consequence of this, the fact that during the past emissions in the South have been much lower than in the North is ignored and also the corresponding question of whether the South should be compensated for this, as some have argued (Ghosh, 1993). Furthermore, intertemporal equity trading might be a reasonable device if the transition from an unfair to a fair allocation involves a sharp increase in the efforts of some agents. An analysis of those dynamic equity issues is certainly a valuable area of further research.

Finally, the application of fair division criteria requires the preceding specification of entitlements to a common property resource – for instance, to a particular service of the environment such as its absorptive capacity for greenhouse gases. In inheritance problems, which are sometimes used to illustrate the theory of fair division, entitlements may indeed be exogenously given through the will of the deceased. However, this is not the case for climate change and the specification of entitlements involves some substantial ethical judgements. In this respect, equal per capita entitlements, which correspond to the justice principle of 'equality of resources', have received particular attention. This is also the case for climate change, where the environment's absorptive capacity for greenhouse gases is often regarded as a global common, as if it were 'manna fallen from heaven'; accordingly, equal per capita entitlements is the proposal mentioned most often in the literature (IPCC, 1996:106; Shue, 1995).

The main objection against this allocation is that even though it might be equitable from an *ex ante* perspective, *ex post* it would lead to an unfair distribution of emission reduction burdens. With tradeable equal per capita permits, the South could sell a large number of permits to the North, which would result in substantial North–South resource transfers (Grubb et al, 1992). Without tradeable permits, the additional abatement cost in the North would be even higher than those transfers, due to the lower efficiency of abatement measures. In arguing this, most writers assume that the initial allocation of emission rights would quasi automatically determine the allocation of abatement cost. However, this need not be the case if one allows for the possibility of additional transfer payments. Accordingly, the main focus in the rest of this chapter is on the fair division of the gains from the exchange of the initial permit allocation, which arise from differences in marginal abatement cost across countries.

---

6 The IPCC report states that 'there are few, if any, ethical systems in which it is acceptable for one individual knowingly to inflict potentially serious harm on another and not accept any responsibility in helping or compensating the victim or to pay in some other way' (IPCC 1996: 101). This would imply that, in view of the unequal distribution of historical and current emissions, the cost of climate change damages would primarily have to be borne by the North.

The Choice of Equity Criteria

The following axiomatic approach to the fair division of common property resources is based on an ordinal utility concept as is commonly used in economic theory. Accordingly, a distribution's equitability is not judged from the allocation of goods but from the allocation of the utility derived from the consumption of these goods. There are five fundamental criteria which will be used to judge the equitability of an allocation mechanism for common property resources.

Efficiency

An allocation is said to be efficient or Pareto optimal if no individual can be made better off without making another individual worse off. Sometimes this has been termed as the criterion of *unanimity*. This already shows that the Pareto criterion is no more than a smallest common denominator, the only normative argument on which the economic profession could agree.

If lump sum transfers are feasible, Pareto efficiency implies that emission reductions have to be allocated such that corresponding marginal costs are equalized in all countries.[7] This is a far reaching result because it determines the allocation of emission reductions. However, it does not say who should bear the cost. After the cake has been maximized, it now has to be shared – as equitably as possible.

The no envy criterion

Probably the most prominent equity criterion within the economic profession is the no envy test (Foley, 1967); indeed, fairness has sometimes been defined as envy freeness plus efficiency (Varian, 1974). With equal entitlements to a common resource, an allocation is said to be envy free if each agent values his share at least as much as anybody else's share. Take the most simple example of parents who bequeath their fortune – consisting of two (identical) apples and bananas – to their two children in equal shares. If one of them likes apples and the other one bananas, they are free to exchange their entitlement to the inheritance. The no envy criterion requires that after this reallocation every heir should be satisfied with his share and not feel the desire to change it against the share of the other heir.

However, the set of envy-free allocations of rights to pollute the atmosphere and the corresponding compensatory payments are quite large, at least if one simplifies the allocation problem so that there are only two agents. There are different allocations where the North compensates the South for a greater share of per capita permits, while neither of them would prefer the per capita permits and compensatory payments of the other one.

---

[7] It has been pointed out in the second section that the equation of Pareto efficiency with equal marginal abatement cost across countries depends on the possibility of lump-sum transfers, which will be assumed to exist throughout the rest of this chapter (see Chichilnisky and Heal, 1994; see also Chapter 3 by Linnerooth-Bayer in this book).

The fair share guaranteed criterion

The fair share guaranteed criterion has been introduced by Steinhaus (1948) and it expresses the idea that every agent should be guaranteed at least the utility from consuming its fair share: its entitlement to the common property resource. This means that if there are overall gains from a reallocation of the initial shares, everyone should be weakly better off after this reallocation has taken place. Other common names for this criterion are *individual rationality* or *acceptability* because one will usually not agree to a reallocation if one becomes worse off than in the starting position.

Applied to climate change, the fair share guaranteed criterion guarantees each country at every time point the utility from the consumption of its fair share of the atmosphere's sink capacity. This can be stated formally as:

$$U_i(P_i) + M_i \geq U_i(FP_i) \tag{4}$$

With $U_i(P_i)$ signifying country i's utility from the permits it receives in the efficient allocation, each country has to be compensated by monetary transfers M_i such that it is at least as well off as with its fair share of permits $U_i(FP_i)$.[8] It should be noted that the fair share guaranteed criterion does not determine what constitutes a 'fair share' of the common property resource, but rather states the implications that follow from a particular specification of fair shares or entitlements.

It can be argued that the utility which an actor receives from a particular number of permits is nothing else but the cost of abatement measures he or she would have to undertake without those permits. Therefore, equation 4 can be rewritten as:

$$M_i \geq C_i(P_i) - C_i(FP_i) \tag{5}$$

where $C_i(P_i)$ and $C_i(FP_i)$ are the abatement cost to reduce emissions from the business-as-usual path to the levels P_i and FP_i respectively.[9] In summary, the fair share guaranteed criterion requires that those countries which receive less than their fair share of permits in the efficient allocation have to be fully compensated for their additional abatement cost. In contrast, those countries which receive more permits than their fair share in the efficient allocation need not pay a higher compensation than the savings in abatement cost that accrue from the permits they receive on top of their fair share. In a nutshell, no country should lose on the way from the original to the efficient allocation.

---

8 Technically, it is assumed that compensatory payments are feasible via a single good (money), in which utility is linear. This representation of preferences by a quasilinear utility function follows from the assumption that the absorptive capacity is given exogenously, and each agent's demand for a share of it depends only on its relative price – ie whether it is cheaper than the emission reductions required otherwise – but not on available income.

9 $U_i(P_i) = C_i(0) - C_i(P_i)$ and $U_i(FP_i) = C_i(0) - C_i(FP_i)$.

The resource and population monotonicity criteria

The criteria of resource and population monotonicity set some limits on how agents' individual utility levels should respond to changes of the allocation problem with respect to the size of the common property resource to be allocated and the number of claimants.

Resource monotonicity requires that if the common resource – in our case the global number of emission rights P_G – grows from P_G to P_G', each agent should be at least as well off as from the fair division of the smaller resource (Roemer, 1986):

$$P_G' \geq P_G \Rightarrow U_i(P_i') + M_i' \geq U_i(P_i) + M_i \tag{6}$$

In climate change and in many other environmental problems, this criterion may indeed have high political relevance. Our best assessment of the environment's absorptive capacity is only preliminary and likely to change as the scientific knowledge improves. Furthermore, reduction targets will often be approached only stepwise. In both cases, the size of the common resource to be divided changes and, due to the commonality of ownership, this should affect the welfare of all agents in the same direction.

The criterion of *population monotonicity* stipulates that if the number of agents entitled to the common resource increases from N to N', no agent should be better off than before (Chichilnisky and Thomson, 1987):

$$N' \geq N \Rightarrow U_i(P_i') + M_i' \leq U_i(P_i) + M_i \tag{7}$$

Similar to resource monotonicity, the criterion of population monotonicity is also based on the ethical argument that common ownership implies a minimum degree of solidarity, namely that everyone should contribute to satisfy the legitimate claims of newcomers.

The stand alone criterion

The monotonicity axioms can be used to deduce the stand alone utility $U_i(P_G)$ – that is, an agent's utility from the consumption of the whole common resource P_G (Moulin, 1992) as an upper bound on the utility an agent may receive from the fair division of a common property resource. To see this, imagine that there is only one country. It would receive its stand alone utility by definition. Population monotonicity requires that the utility of this country does not increase as the number of countries with legitimate claims to the common resource increases – in other words, it may not receive more than its stand alone utility. The same conclusion can be derived from the criterion of resource monotonicity. Applied to climate change, this can be stated formally as:

$$U_i(P_i) + M_i' \leq U_i(P_G) \tag{8}$$

Following the specification of utility functions as introduced above, this is equivalent to:

$$M_i \leq C_i(P_i) - C_i(P_G) \tag{9}$$

While the fair share guaranteed test fixed a lower bound to the compensatory payments, the stand alone test sets an upper bound, namely that no country will receive a higher compensation M_i than the abatement cost it would save with the global number of permits P_G.

More generally, it follows from the monotonicity criterion that no agent should be better off when the environment's absorptive capacity for pollutive emissions is a scarce resource compared to the case where the environmental problem does not exist. Moulin (1992:1333) justifies the stand alone test by arguing that 'fair division conveys the idea of no subsidization: the presence of other agents who are willing to pay higher monetary transfers than me for consuming the resources should not turn to my advantage'. This argument seems particularly justified if the willingness to pay higher monetary transfers is related to efforts to reduce a problem which affects all agents – such as climate change. In this context one could state the stand alone test bluntly as: 'no one should benefit from the abatement burdens of others', which reflects the solidarity idea underlying the monotonicity axioms.

Towards a Fair Division of Climate Protection Burdens

Taken for itself, the fair share guaranteed and stand alone criteria sound rather mild, but their combination leads to far reaching results. As long as the South's fair share of emission rights is higher than its business-as-usual emissions, it will not have to undertake any abatement efforts to reduce emissions to the level of its fair share of permits: $C_S(FP_S) = 0$. If one assumes realistically that global emission budgets are reduced only stepwise, this will be the case if entitlements are specified in equal per capita terms, but it suffices that the South's entitlements are not much lower than this. Because a country's fair share FP_i is always smaller or equal to the global number of permits P_G, it follows from $C_S(FP_S) = 0$ that $C_S(P_G) = 0$. Therefore, the fair share guaranteed criterion (equation 5) and the stand alone criterion (equation 9) can be combined to yield:

$$C_S(P_S) \leq M_S \leq C_S(P_S) \text{ or } M_S = C_S(P_S) \tag{10}$$

This means that as long as business-as-usual emissions in the South are lower than its entitlements, there exists only a single solution that is efficient and which satisfies the fair share guaranteed and the stand alone criteria. The North will have to pay the full incremental cost of emission reductions in the South – but not more. On the other hand, the South will have to agree to an efficient allocation of abatement measures. It should be noted that this solution is also envy free, because the South would not prefer the permits and compensatory payments of the North or vice versa (Helm, 1998).

So far, attention has been restricted to the first phase of a climate protection regime where the South's fair share of permits is greater than its business-as-usual emissions. At least a short look will be taken at later periods.

Within the economic theory of justice, the distribution of divisible common property resources via the assignment of property rights, and a subsequent allocation via competitive markets, has a very prominent position (Young, 1994:161). This mechanism does not only lead to an efficient allocation, but it also satisfies the equity criteria of fair share guaranteed and envy freeness (Moulin, 1990). However, due to the initially very large differences in per capita emissions and marginal abatement cost in the South and North, the competitive allocation would violate the stand alone test and the criteria of resource and population monotonicity respectively. Nevertheless, the market solution still constitutes a primary focal point as soon as per capita emissions and abatement cost in the North and South approach each other. But when exactly should the transition to this allocation mechanism take place?

If it begins immediately after the first period – when the South's fair share of permits equals its business as usual emissions – the South can still sell permits to the North until marginal abatement costs are equalized globally. Because with competitive markets the permit price would be equal to the (increasing) marginal cost of the last unit reduced, a profit similar to the producers' surplus would accrue to the South. This again would be a violation of the stand alone test.

This suggests a solution in which the South receives the minimum of the competitive allocation and the stand alone value. The transition from the first solution to the second would roughly take place when those abatement costs in the South, that result from the necessity to share the atmosphere's sink capacity with the North, are equal to the North–South transfers in the market allocation. The North, on the other hand, would receive its share of the competitive allocation, including the money that the South would receive on top of its stand alone utility in the competitive allocation.[10]

CONCLUSION

Starting from some general criteria for an equitable solution, a precise proposal has been developed for allocating emissions and abatement costs in a climate protection regime. In particular, it has been suggested that the South should initially be fully compensated for the abatement cost it has to undertake in order to foster the efficiency of emission reductions. In many respects, this solution is surprisingly similar to the international agreement to combat stratospheric ozone depletion, which has been praised for its fairness. In this case, low emission countries have been fully compensated for their incremental abatement cost by high emission countries (see Biermann, 1995).

Unfortunately, exact implementation of the proposed solution is not straightforward, because it requires information on countries' abatement costs, which serve as a basis for compensatory payments. Yet, from a political point of view, this can also be seen as an opportunity since it opens a certain leeway for negotiators. Perhaps the most obvious procedure would be to start with an initial phase of joint implementation, which would later be succeeded by a tradeable permit system. Interestingly, this is not so different from the provi-

---

10 For a general formulation of this allocation mechanism with more than two agents, see Helm (1998).

sions in the Kyoto Protocol which provide for a permit system among industrialized countries and a system similar to joint implementation – the clean development mechanism – between industrialized and developing countries.

Certainly, not everyone will perceive this proposal as equitable. This might be due, partly, to the decision of separating the distribution in the climate protection regime from the higher level of the global welfare distribution. This was necessary for the chosen axiomatic approach, and the author believes that in the present case the breaking down of the problem into its individual parts can improve the view on the whole. But one might postulate that the South's share, as determined above, should be regarded as a minimum, and if global welfare differences were taken into account it should, rather, receive more.

Similarly, some might not agree with the equity criteria themselves. In this case, the axiomatic approach could at least help to focus discussions on those disputed criteria. However, perhaps it has become clear that something more scientifically can be said about equity than that it is 'simply a matter of the length of the judge's ears', as Elbert Hubbard once put it provocatively.

REFERENCES

Biermann, F, 1995, *Saving the Atmosphere: International Law, Developing Countries and Air Pollution*, Peter Lang Verlag, Frankfurt am Main, Germany

Brams, S J and Taylor, A D, 1996, *Fair Division: From Cake Cutting to Dispute Resolution*, Cambridge University Press, Cambridge, MA

Chichilnisky, G and Heal, G, 1994, Who should abate carbon emissions? An international viewpoint, *Economic Letters* 44:443–449

Chichilnisky, G and Thomson, W, 1987, The Walrasian mechanism from equal division is not monotonic with respect to variations in the number of consumers, *Journal of Public Economics* 32:119–124

Cline, W, 1992, *The Economics of Global Warming*, Institute for International Economics, Washington, DC

Foley, D, 1967, Resource allocation in the public sector, *Yale Economic Essays* 7(1):45–98

Ghosh, P, 1993, Structuring the equity issue in climate change, in: Achanta, A N, ed, *The Climate Change Agenda: An Indian Perspective*, Tata Energy Research Institute, New Delhi, India, pp 267–274

Grubb, M, Sebenius, J, Magalhaes, A and Subak, S, 1992, Sharing the burden, in: I M Mintzer, ed, *Confronting Climate Change: Risks, Implications, and Responses*, Cambridge University Press, Cambridge, pp 305–322

Harsanyi, J, 1955, Cardinal welfare, individualistic ethics, and interpersonal comparisons of utility, *Journal of Political Economy* 63:309–321

Helm, C, 1998, *A general mechanism for the fair division of common property resources with an application to climate change*, Paper presented at the European Economic Association Congress, 2–5 September, 1998, Berlin, Germany

IPCC (Intergovernmental Panel on Climate Change), 1996, *Climate Change 1995: Economic and Social Dimensions of Climate Change*, Contribution of Working Group III to the Second Assessment Report of the IPCC, Cambridge University Press, Cambridge, UK

Kolm, S-C, 1997, *Modern Theories of Justice*, MIT Press, Cambridge, MA

Linerooth-Bayer, J, Climate change and multiple views of fairness (Chapter 3 of this book)

Moulin, H, 1990, Fair division under joint ownership, *Social Choice and Welfare* 7:149–170

Moulin, H, 1992, An application of the shapley value to fair division with money, *Econometrica* 60(6):1331–1349

Moulin, H, 1995, *Cooperative Microeconomics: A Game-Theoretic Introduction*, Princeton University Press, Princeton

Rawls, J, 1971, *A Theory of Justice*, Harvard University Press, Cambridge, MA, USA

Rawls, J, 1993, The law of peoples, in: S Shute and S Hurley, eds, *On Human Rights: The Oxford Amnesty Lectures 1993*, Basic Books, New York, pp 41–82

Roemer, J E, 1986, The mismarriage of bargaining theory and distributive justice, *Ethics* 97:88–110

Roemer, J E, 1996, *Theories of Distributive Justice*, Harvard University Press, Cambridge, MA

Shue, H, 1995, Ethics, the environment and the changing international order, *International Affairs* 71(3):453–461

Simonis, U E, 1996, Internationally tradeable emission certificates: Linking environmental protection and development, *Economics* 53:96–110

Steinhaus, H, 1948, The problem of fair division, *Econometrica* 16(1):101–104

Varian, H R, 1974, Equity, envy, and efficiency, *Journal of Economic Theory* 9:63–91

Young, H P, 1994, *Equity in Theory and Practice*, Princeton University Press, Princeton

6 THE APPROPRIATENESS OF ECONOMIC APPROACHES TO THE ANALYSIS OF BURDEN-SHARING

H Asbjørn Aaheim

INTRODUCTION

Burden-sharing has become a catchword for issues related to distributing commitments of climate policy targets among nations. The relation between distribution and principles of fairness is, no doubt, an Achilles heel in economic analysis. Economists feel more comfortable when they can consider the distribution as given, or exogenous, and concentrate on the allocation of resources. As pointed out in other chapters of this book, classical economics therefore usually confines itself to studies of Pareto optimality (see Chapters 2 and 5), and is highly insufficient for a full examination of the controversial issues of fairness that inevitably emerge in climate negotiations. One may also fear that some economists overemphasise the importance of efficiency and effectiveness at the expense of the more controversial issues of fairness and distribution. An unfortunate consequence is that people sceptical of economics, of whom there are many, disregard economic arguments.

On the other hand, effectiveness and efficiency are important aspects when designing a climate treaty, also in the context of burden-sharing. One example is that a cost-effective implementation of a climate treaty implies that there is more to distribute, compared with a non-effective treaty. Perhaps more important is that in order for a climate treaty to be cost effective, the marginal cost of reducing greenhouse gas emissions ought to be the same across countries. This coheres nicely with the polluter pays principle, which is appealing in the context of fairness. Everything else being equal, there is no reason why people in one country should pay more for a climate treaty than people in another country. The difficulties arise, however, when everything else is unequal. This indicates that economics may be used to address some important issues raised in the debate about burden-sharing, but the implications should be interpreted with caution.

This chapter has three major aims. Firstly, it shows what economic analyses of climate policy have contributed in the context of distribution and fairness. Secondly, it points at some aspects of burden sharing, often ignored by economists, that could be addressed adequately with available analytical tools. Finally, it discusses some issues that require a deeper theoretical examination before they can be addressed adequately by economic analyses.

Some vitally important issues concerning fair distribution are clearly beyond the scope of economic analysis, as we understand it today. Rayner et al (see Chapter 2) distinguish between intentional and behavioural approaches, and economics is basically founded on the behavioural approach. Many issues in intentional analyses are difficult to analyse with a behavioural approach without losing important information. For instance, we know of many factors that can explain changes in individual preferences. Such a 'mapping' is interesting and fruitful for many reasons. Some of these factors may be systematized in order to establish preference maps and to carry out behavioural analyses, but it is a waste of resources to try to include all possible factors in preference functions.

CONTEXTS IN WHICH BURDEN-SHARING CAN BE DEFINED

It is not clear what is usually meant by burden-sharing in the context of climate policy. Sometimes burden can be interpreted as any kind of resistance resulting from initiatives taken to mitigate climate change, for instance political resistance against carbon charges or in terms of measurable economic costs. In other contexts, burden relates to the effects of global warming. In this paper, we restrict the use of burden to economic costs, but define costs in a broad sense. The concept of burden will, however, be used with different content depending upon the context in which it is used in economic analysis. It may be useful to distinguish between burden-sharing in analysis of cost effectiveness, of cost efficiency and in welfare analysis.

Cost effectiveness has been the major subject for economic analysis of climate policy until now. The point of departure is that the maximum level of emissions, the emission target, is given by default, and the aim is to achieve the target at a minimum cost. Fairness is reflected in these studies in the sense that a cost-effective implementation of a given target requires that the marginal cost of emission reductions is equalized across countries. Therefore, all the polluters pay the same and the full price for the achievement – the polluter pays principle. In the context of cost effectiveness, burden relates to the distribution of abatement costs.

Cost efficiency means that the abatement costs are considered with reference to the expected benefits of reducing the emissions of greenhouse gases. For the purpose of this paper, we use the concept of cost efficiency with reference to evaluations on the national level. Cost efficiency differs from cost effectiveness because under efficiency one also asks how much abatement it is worthwhile carrying out. Consequently, abatement costs are regarded as a means to gain benefits. In this context, it is reasonable to define burden as the net cost of a climate treaty (benefits minus costs), which may be positive, zero or negative. Fairness in this context is clearly very different from the polluter pays principle following cost effectiveness, since it takes into account who gains the most for concerted action against climate change. The ideal treaty is one where all countries end up with a zero net cost. In most cases, however, such a treaty is unfeasible. Analysis of cost efficiency as applied in this chapter is particularly appropriate when analysing the climate negotiations.

Welfare analysis can be applied to normative studies of what 'the world' ought to do in order to mitigate climate change. It extends the perspective of cost-efficiency analysis by reference to an explicit definition of fairness. As already pointed out, it makes economists uneasy to be explicit about the fairness of distribution. Cost-benefit analyses of optimal global climate policy that have been carried out (see Manne et al, 1994; Nordhaus, 1993; Cline, 1992) therefore take the existing 'observed' distribution of income between countries as the point of reference. As one may expect, this has led to a vigorous debate about the use of economic tools for analysing climate policy.

With reference to the aims of this chapter, it is fair to say that economics have contributed a lot to the analysis of cost effectiveness. There is a potential for economic analysis of cost efficiency, but there has been a limited activity in this field. Economic welfare theory will probably have to be elaborated upon further in order to analyse the concerns for a fair global distribution of commitments in climate policy. In the remainder of this chapter, the strengths and weaknesses of the economic approaches to burden-sharing are discussed in more detail.

COST EFFECTIVENESS

Some individuals seem to deny that criteria for cost effectiveness are relevant for assessing a fair distribution of commitments in a climate treaty. Nevertheless, cost effectiveness was a major issue prior to the Kyoto meeting in December 1997, when the first binding commitments to reduce greenhouse gas emissions were negotiated. This was partly reflected in the views on how the cuts were to be distributed. Most countries considered flat rate reductions to be the only feasible key for distributing commitments among the industrialized countries.[1] Some countries, notably Australia, Canada, Japan, Iceland and Norway, held the view that flat rate reductions are ineffective because the marginal cost of abatement varies among countries. In terms of the polluter pays principle, some polluters therefore have to pay more than others, which these countries considered to be unfair. In many cases, they argued, the highest cost will be imposed on the countries with the lowest emissions per capita. Hence, countries are being punished for being 'clean'.

There are many studies of the national costs of emission cuts that may serve as a basis for differentiating between commitments among countries. The studies fall into two main categories. One category, bottom-up studies, estimates the costs of implementing new technology with lower carbon emissions in different countries. The other approach is based on macroeconomic models (top-down studies), and estimates how alternative targets for emission reductions implemented by carbon charges affect the gross national product, or GDP. Figure 6.1 shows a comparison of the estimated costs of emission reductions by

---

1 By flat rates, all countries reduce their emissions by the same per cent relative to a base year.

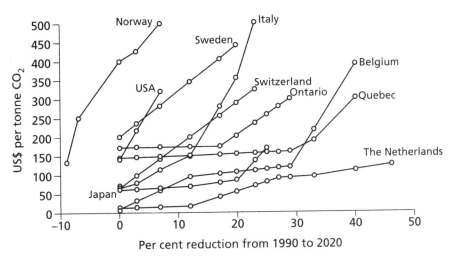

Source: Kram and Hill, 1996

Figure 6.1 ETSAP estimates of national costs of alternative emissions reductions in 2010

new available technologies in some OECD countries. The results are based on bottom-up studies within the Energy Technology Systems Analysis Programme (ETSAP) network (Kram and Hill, 1996). The figure shows significant differences but contains some peculiar results: the US exhibits high costs, while the costs in Japan are considered to be relatively low. Moreover, the study expects The Netherlands to be able to reduce its emissions by 50 per cent with lower unit costs than the US would have to pay for stabilization.

The results illustrate the problems of establishing uniform standards for the assumptions in this kind of studies. Ideally, one should study implementing the technology frontier in all countries on a large scale – for instance, requiring the same energy efficiency in fossil fuel power plants. However, the opportunities in one country may differ substantially from the opportunities in another country because the potential for emission reductions varies across sectors. It is very difficult to define equal standards across sectors – for instance, to compare standards for power plants with standards in the transport sector. The setting of standards will easily differ across sectors with respect to the possibility of implementing the technologies. The reality of the technology assumptions in one country may therefore be very different from the reality of the assumptions in another country with a widely different composition of sectors.[2]

Another problem relates to the measurement of costs. If greenhouse gas emissions are to be reduced significantly, this will have a substantial impact on the energy market and thereby affect energy prices. Bottom-up studies are, however, usually based on exogenous prices. Moreover, the amount of emission reductions depends on the expected economic growth rate, which differs among countries. Hence, bottom-up studies give insufficient information about the required abatement a country faces when reducing their emissions by a certain per cent compared to a base year.

---

2 A similar ETSAP study pays particular attention to this problem.

As an alternative, estimates of the national costs could be based on alternative runs of a macroeconomic model. It is not straightforward to define the national cost in a macroeconomic context. Most analysts use the reduction of GDP following a policy to reduce the emissions of CO_2. Weitzman (1976) gives a justification for using GDP as a measure of welfare, but only within a highly stylized macroeconomic model. The result is based on a comparison of two consumption paths from now until infinity: a constant path and an optimal consumption path starting from the present consumption level. He shows that these two paths yield exactly the same total welfare if the level of the constant consumption path is equal to the present GDP. Some justification for using GDP is thereby provided, although the result does not apply if other, perhaps more reasonable, assumptions are being used (Asheim, 1994; Brekke, 1994). Supplementary information, such as the required carbon tax to achieve a given reduction in emissions, is therefore useful and relevant when considering the costs in terms of the loss of GDP.

Studies of the reductions in GDP following targets for reducing CO_2 emissions have been made in a number of countries. Results from some of these studies are shown in Figure 6.2, which is based on IPCC (1996). A major reason why GDP is affected differently in different countries is the economic structure and the structure of the energy demand. The differences may be due to a varying sector composition. Of particular importance is the size of the energy intensive industry and the energy resources on which the electricity sector is based. Differences in sector composition among countries may, to some extent, be considered as a result of the international division of labour. Countries with a large export of carbon intensive products will have to pay a relatively high price for a given per cent reduction in their CO_2 emissions than countries with a low carbon content in their export products. One may therefore argue, as did Australia before the Kyoto meeting, that the carbon content of all commodities should be charged for emissions in their final use.

One might expect that these differences would be demonstrated when comparing the cost of CO_2 abatement in macroeconomic studies of different countries. As we can see from Figure 6.2, such a comparison is difficult to carry out. This is partly due to the different structures of the models and to different assumptions in the studies. The effects of climate policy strategies are highly sensitive to assumptions about the effects on foreign trade, beliefs about technological change, etc. In addition, macroeconomic models do not specify technological options to reduce emissions, since the technologies are represented by relations between aggregated input and output. As a consequence, it is usually assumed that all measures to reduce emissions are invoked by carbon charges. This is reasonable in the case of CO_2 emissions, but insufficient if we include other possible options in climate policy, such as carbon sequestration and abatement of other greenhouse gases than CO_2. For instance, Aaheim (1997) shows that if Norway is to cut its emissions of greenhouse gases by 20 per cent, it is cost effective to cut 10 per cent through charges, and 10 per cent through other means.

Macroeconomic country studies apply for assessing the level of climate action within single countries, but they are difficult to use for comparing the cost of emission cuts among countries. In particular, effects of a concerted action against global warming on the world market affect countries differently (Torvanger et al, 1996; DFAT and ABARE, 1995). Moreover, flexible implemen-

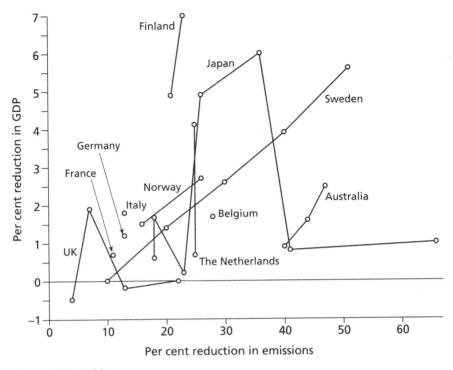

Source: IPCC, 1996

Figure 6.2 National costs of CO_2 emission reductions according to national top-down studies

tation of the agreement will have a greater impact on some countries than on others. Flexible mechanisms mean that one country can buy emission cuts in other countries, either by buying emission allowances (tradeable quotas) or by investing in measures to reduce emissions in other countries (activities implemented jointly and clean development mechanisms). The aim of flexible mechanisms is to equalize the user cost of emitting greenhouse gases, and thereby achieve cost effectiveness. Macroeconomic analysis is particularly appropriate for analysing these mechanisms. The studies are founded on a well-known theoretical basis with a limited influence of value judgements, and they focus on the allocation of resources, which is considered by many economists as the central subject for economic analysis.

How flexible mechanisms may lead to a 'fair' redistribution of emission targets can be illustrated in a simple figure (see Figure 6.3). The figure displays the curves for the marginal cost of abating greenhouse gas emissions for two countries, A and B. A's marginal cost of emission reductions are increasing from left to right in the diagram, and B's marginal cost of emission reductions are increasing from right to left. Assume that the initial quota implies an emission reduction indicated by the thick, dashed vertical line *Q*. In this point, the marginal cost of A's abatement is *c(A)* and B's is *c(B)*. Hence B would gain if it could achieve emission reductions at a lower cost than *c(B)*. A, on the other hand, would be willing to abate more if it was compensated by at least as much as *c(A)*. Thus, A and B have motives to trade with each other until A's

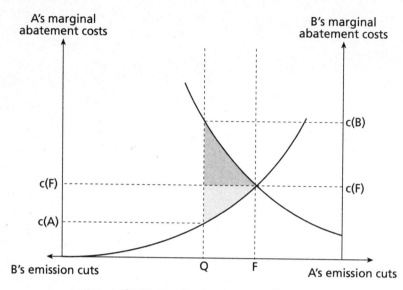

Figure 6.3 Redistribution by cost-effectiveness

marginal cost is $c(F)$, which becomes the equilibrium price of emission allowances in a flexible regime.[3] The traded amount between the two countries is $F - Q$. A's benefit from this trade equals the light shaded triangle, and B's benefit is the dark triangle.

Flexibility reduces the marginal costs of those countries that face high abatement costs by levelling out cost differences. One may, therefore, say that flexible mechanisms contribute to strengthen the fairness of a given distribution of quotas. However, the problem of distributing quotas is not solved for that reason. Clearly, the country that is able to sell emission allowances is better off than the country that has to buy. One could argue that this is fair if one regards low-cost abatement opportunities as a kind of wealth to a nation, just like natural resources, which should give a certain return. Then the benefits gained by A above are reasonable. The argument used by countries with high abatement costs referred to above, however, goes against this.

COST EFFICIENCY

The fairness in equalizing marginal abatement costs across countries relates to the perception that everyone should pay the same price for the same good. One may, however, question whether it is correct to regard a given reduction in emissions to mitigate global warming as the 'same good' for all countries. What if country B considered global warming to be a serious threat, while country A did not pay too much attention to it, for instance because it considered the advantages to be more important than the disadvantages. Then A's benefits from flexible mechanisms could not compensate the loss they have to

---

[3] Note, however, that with two parties it is unlikely that this competitive equilibrium will be realized.

bear because of an unfavourable treaty. B, on the other hand, might be willing to pay more than $c(B)$ to get the treaty that distributed the quota Q. As a result, A turns out to be the loser, while B is the winner.

From this perspective, it might be tempting to evaluate burden-sharing by considering the *net cost* of climate policy commitments (benefits from reduced global warming minus the costs of abatement and adaptation). The benefits may partly be based on estimates of the cost of the expected damage from global warming, which differs from country to country, but they may also vary because of differences in the willingness to pay for a given damage. To evaluate the distribution of net costs of a negotiated treaty, we need to assess both the damage costs and the willingness to pay.[4] This is clearly controversial because the willingness to pay is closely dependent upon the ability to pay. A comparison of net costs across countries is, therefore, an inappropriate point of reference for assessing the distribution of commitments among countries of widely different affluence. However, it may be acceptable for comparing relatively equal countries, and of great interest for understanding the varying interests and positions of different countries.

Economists use the term efficiency to denote the point at which a cost-effective solution also maximizes the benefits, as distinguished from effectiveness, where the benefits are not subject to control. In global matters, efficiency can be interpreted in two ways. Efficiency on the national level means that each country considers its own abatement costs in light of what it expects to gain in terms of mitigated global warming. Efficiency on the global level must relate to a unified concept of global welfare, in which principles of fairness across countries and groups of individuals are embedded. In this section, we concentrate on the national level. Relations to principles of fairness will be discussed in the next section.

A number of assessments of damage costs from global warming have emerged in recent years. Fankhauser (1995) summarizes and compares some of them. This is an immature field of economic analysis and the estimates differ considerably, besides being highly uncertain. The estimates refer to concentration levels at a doubling of CO_2 ($2 \times CO_2$) in the atmosphere, but the authors use different assumptions about the temperature rise. There are also huge differences in the estimates of varying damage components. Titus (1992) estimates the damage to forests in the US to be US$38 billion, while Cline's (1992) estimate is US$2.9 billion. Tol (1993) evaluates the loss of human amenity in the US as US$12 billion, while other authors disregard this aspect. A particularly sensitive question concerns estimates on mortality and morbidity, for which the authors differ in the range of 1:8. For the US in total, the estimated cost varies from US$49 billion (Nordhaus, 1991) to US$121.3 billion (Titus, 1992), which constitutes between 1 and 2.5 per cent of GDP.

Fankhauser's (1995) estimates of the damage in world regions vary from 0.7 per cent of GDP in ex-USSR to 4.7 per cent in China. With reference to the net cost, one may therefore expect that countries have widely different interests in pursuing a global climate policy. Torvanger et al (1996) show how different abatement and damage costs may affect the interests of a country.

---

4 Aaheim et al (1998) discuss the relations between damage cost estimates and estimates of the willingness to pay.

Table 6.1 Assumptions about abatement costs and damage costs (per cent of GDP)

	Abatement costs		Damage costs
	10% reduction in CO_2 emissions	30% reduction in CO_2 emissions	2 x CO_2
Country A	0.5	3.0	2.0
Country B	0.5	3.0	4.5
Country C	0.1	1.5	2.0

Consider three countries that differ only with respect to damage costs and abatement costs. Assume also that they base their interests on a comparison of abatement costs and damage costs under equal participation of a climate treaty, or 'flat' rates of emission reductions for all countries of the world. That is, x per cent reduction in national emissions means that global emissions are reduced by x per cent. More specifically, we use the assumptions displayed in Table 6.1.

The abatement costs should be interpreted as long-term costs. The damage costs in countries A and C are somewhat more pessimistic than the average estimates, while the damage costs in country B are in the upper level of the interval. The level of abatement costs in A and B is approximately the same as estimated long-term costs for low levels, but higher for a high level of abatement (see IPCC, 1996). The abatement costs in C are low compared with most studies of industrialized countries. However, these studies usually disregard secondary benefits of CO_2 emission control. This may cause a considerable overestimation of abatement costs (see Alfsen et al, 1992; Ekins, 1995; Aaheim et al, 1997).

We can use an aggregated, intertemporal growth model to assess the optimal emission path for each country. The model maximizes total welfare over a given period, where the welfare depends exclusively on consumption. The economic product depends on the capital stock and the concentrations of CO_2, which are assumed to be indicators for climate change. Hence, the national product goes down if the concentration of CO_2 increases. The national product is allocated to three purposes: consumption; investments; and abatement of greenhouse gases. Emissions are assumed to be proportional to output. A certain rate of technical change is implied, which indicates that the energy intensity is reduced over time.

The model gives conditions for how consumption, investments and abatement costs should be allocated from now on until a given future year in order to maximize total welfare. This depends upon the requirements set on the stock of real capital and the concentrations of greenhouse gases at the terminal year of the period. If no such requirements are given, it is clearly optimal, for instance, to 'consume' the whole capital stock before the end of the period. To account for the generations to come after the terminal year, we require, firstly, that the wealth of the capital stock is at least as large at the terminal year as it is today. Secondly, it is assumed that the negative wealth (debt) of CO_2 concentrations is no greater at the terminal year than it is today. Wealth and debt are defined as physical stock multiplied by the shadow prices, which are

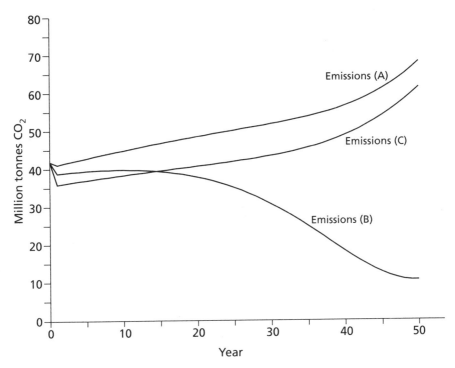

Figure 6.4 Optimal emissions of CO_2 in cases A, B and C

solved in the model. In the terminology of Asheim (1993), this means that we require strong sustainability.[5]

Figure 6.4 shows optimal emission paths for countries A, B and C. The interests differ considerably with respect to emission paths. Country A will encourage a slight increase in emissions from year 0. Country B wants to reduce emissions slightly, and argues that the emissions should be stabilized on that level for about 20 years. After that, emissions should decrease significantly and approach one quarter of present emissions in 50 years. Country C may agree with B that emissions should be stabilized, but only within a 20-year perspective, since it is in favour of an immediate reduction of more than 10 per cent. This opinion is due to C's low abatement costs. If negotiating about targets 15 years ahead, B's and C's opinions concur, but the long-term interests diverge radically. In the long term, C will encourage growth in emissions. B seeks to mitigate an increase in CO_2 concentrations due to their high damage costs. In the long term, C tends to agree with A.

The net cost of a climate treaty could be defined as the loss that each country faces if it has to follow a treaty that diverges from its 'best' or most preferred one. This loss can be divided into two components. One component stems from the diverging consumption paths and the other from the fact that the 'debt', or terminal value, of CO_2 concentrations differs in the two cases. Both components can be positive or negative. If B succeeds in achieving the

---

5 Weak sustainability means that *the total wealth* of capital and CO_2 concentrations is the same at $t = T$ as it was at $t = 0$.

treaty it considers to be the best, the costs in A and C are a result of a low level of consumption, and a low level of CO_2 concentrations in the terminal year compared with their optimal choice. The diverging consumption levels add to the costs, while the low level of CO_2 concentrations reduces the costs. If A succeeds in attaining its best choice, the total net cost imposed on B is due to a high level of concentrations at the end of the period. This adds to the cost. On the other hand, A's best treaty allows B to attain a higher consumption level, which reduces the cost. By definition, however, the total cost is positive; but it is not straightforward to define the total cost. The final outcome of negotiations is, moreover, that all countries will have to give and take. Thus all countries will probably end up with a positive cost, and burden-sharing is then a question of how the net costs were distributed among countries.

In this example, we compare countries that differ only in their abatement and damage costs. In that case, it is acceptable to compare the net costs to evaluate fairness. It is much more difficult to use the net costs as a criterion for a general evaluation of fairness. Recall that the expected benefits of mitigating climate change depend, in principle, also on the willingness to pay. In the example here, the willingness to pay is reflected in the damage cost estimates. Some estimates are based directly on surveys of the willingness to pay – defined by those who were asked. Others are based on observed prices of today and are defined by the preferences of the people participating in the markets. If observed in other countries, or at other periods of time, the estimates could be different. As a consequence, it is highly problematic to use the net costs to make normative assessments of burden-sharing. Some might argue that it is better to compare abatement costs than to compare net costs, which only complicate the matter. The difference is, however, that an evaluation of net costs means that the problems in comparing countries become more transparent. In fact, the use of net cost has a wider application than the use of abatement costs. While a comparison of abatement costs should be restricted to countries on an equal level of development and with approximately the same damage costs, one may expand the applicability of economic comparisons to include countries with different damage costs by using the net costs. In both cases, however, one should limit the comparisons to countries at the same level of development.

GLOBAL WELFARE – BURDEN-SHARING ACCORDING TO JUSTICE

What became transparent in trying to base a fair distribution of commitments by comparing net costs was that it is difficult to base an evaluation of 'fairness' on an aggregate of individual preferences, or of the preferences of several countries. For many years, economists tried to establish welfare functions on the basis of individual preferences. This, they thought, would assist analyses of social choice without any influence of the analyst's norms, just like in the utility theory, where the analysis of the demand can be carried out without a priori statements about the taste and norms of the individuals. Arrow (1951) put an end to these efforts by his famous impossibility theorem. In short, he showed that one cannot aggregate individual preferences into a 'neutral' social welfare function to express the will of the society. As pointed out by Sen

(1979), such a welfare function has to be based on a priori choices about distribution, or so-called interpersonal comparisons.

Arrow used the term 'dictatorship' for a priori choices of distribution. This may have prevented many economists from making explicit normative analyses of distribution. Instead, they have hidden the normative elements in observations of distribution. This is clearly unsatisfactory when analysing the burden-sharing of commitments in climate policy. An integration of moral philosophy and economics is therefore highly appreciated. There is a rapidly growing interest in these problems, but it has not yet penetrated applied economics. One exception is, however, the work of Fankhauser et al (1997), who estimate damage costs of climate change with reference to alternative welfare functions based on different assumptions about the income distribution among countries. They show that the estimated damage cost reported, for instance, by the IPCC (1996) may be significantly higher than previously estimated, if based on a weighing of equity factors between individuals and countries.

A further integration of equity factors clearly represents a challenge for economic analyses of climate policy. Assessing emission cuts based on principles for equity may result in a radically different sharing of commitments than effectiveness and the efficiency criteria indicate. Until now, a number of concepts of fairness and equity applied to the distribution of commitments and responsibility for climate change can be identified. Some of them are summarized in Table 6.2.[6] The list includes the principle of cost effectiveness discussed earlier, which aims to equalize the marginal cost of abatement. 'Flat rate' reductions, which were proposed by the US and the EU, among others, prior to the Kyoto meeting were defended mainly because of their simplicity. They may, however, also be related to the principle of sovereignty, where all countries are considered to be sovereign in their evaluation of the 'baseline'.

Table 6.2 Selected equity principles and examples of related burden-sharing rules

Equity principle	Interpretation	Burden-sharing rule
Polluter pays	All should pay the same amount for each unit of emissions.	Uniform charge
Sovereignty	The present emission level constitutes the basis on which reductions should be distributed.	Flat reductions
Egalitarian	All have the same right to use atmospheric resources, and the same responsibility to preserve them.	Reductions proportional to emissions per capita
Horizontal	All have equal rights to emit the same amount of greenhouse gases to achieve the same level of welfare.	Reductions proportional to emission per GDP
Vertical	Income should be equally distributed among all people.	Reductions proportional to GDP per capita

6 For a closer discussion of the principles, see Rose (1992), DFAT and ABARE (1995), Bureau of Industry Economics (1995), Burtraw and Toman (1992) and Torvanger et al (1996).

This concurs with the view discussed above, where low-cost options were considered to be a kind of wealth to a country. The principle of sovereignty is supported by the Framework Convention on Climate Change, by reference to the Charter of the United Nations (see Torvanger et al, 1996).

While the cost effectiveness and the sovereignty principles relate primarily to an economic point of departure, other principles are more explicitly based on the view that every individual is inviolable, and that all individuals ought to act in accordance with general rules. Hence, it should be acceptable to apply the rules to everyone. These principles are firmly rooted in Western philosophy, such as in Christian ethics, in Kant's moral philosophy and more recently in Rawls's (1971) theory of justice.

The egalitarian principle is based on egalitarian theory, which states that every individual has the same rights and duties, and is therefore a strong expression for the notion of equal rights. Applied to climate change this principle implies that every individual has equal rights to use the atmosphere and to be allowed to emit the same amount of greenhouse gases. Hence, emission permits could be given an equal amount of allowances. An alternative, and weaker, interpretation of the egalitarian principle is to relate it to emission reductions. This would mean that the emission reductions ought to be proportional to emissions per capita.

A distribution according to the net costs as discussed in the previous section may be related to a principle where similar economic circumstances have similar emission rights and burden-sharing responsibilities. This is often referred to as the horizontal principle, since it relates responsibilities to comparably equal countries. Hence, the comparison of abatement costs to net costs could apply if one bases burden-sharing on the horizontal principle. One may think of alternative ways to make the horizontal principle operational to attain a principle for burden-sharing, since we have shown how difficult comparisons of costs turn out in practice. An alternative could be to equalize the cost of abatement in proportion to GDP across countries.

The vertical principle aims at levelling out income inequalities between individuals and countries. Progressive income taxes are a well-known example of how the vertical principle can be implemented. The distinction between the responsibilities of Annex and non-Annex countries is an expression for the impact of vertical principles in the Framework Convention on Climate Change. Hence, low income countries are given less responsibility for mitigating climate change than are high income countries. A rule for burden-sharing in accordance with the vertical principle could be to reduce emissions in proportion to GDP per capita.

To see the practical implications of the egalitarian, the horizontal and the vertical principle, Figure 6.5 shows the estimated percentage reduction in CO_2 emissions for 11 regions of the world if we apply the burden-sharing rules listed in Table 6.2. The egalitarian principle implies that countries with high emissions should reduce more. These countries will often face high reduction costs, since they usually embody an advanced technology with relatively high energy effectiveness. The egalitarian principle will therefore exhibit large cost differences between countries, in a way that the highest costs are imposed on rich countries. For economically equal countries, the principle implies a 'punishment' of countries with low energy effectiveness since these are given the highest reduction.

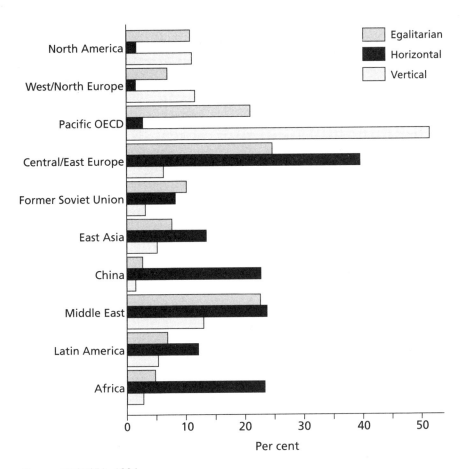

Source: OECD/IEA, 1994

Figure 6.5 Reductions in CO_2 emission by region at a 10 per cent reduction in world emissions according to alternative principles

The horizontal principle is based on the idea that countries should strive to reach a living standard with a minimal load on the global environment. As for the egalitarian principle, countries with low energy effectiveness will face more radical emission reduction targets. The horizontal principle does not, however, lead to higher reductions for rich than for poor countries. On the contrary, countries with low energy effectiveness, which often are the poor ones, will be given the strongest requirements. An additional aspect of this principle follows from the division of labour worldwide. Countries with carbon-intensive export products will have to reduce more of their emissions than other countries, including the countries that actually consume these carbon-intensive products.

By the vertical principle, emissions should be reduced in relation to income; this suggests that rich people should pay more for their emission reductions than poor people. The vertical principle does not, however, give any incentives to enhance energy effectiveness.

The three principles give rise to widely varying requirements for the different regions. In general, the egalitarian and the vertical principles place a higher burden on rich countries, while the horizontal principle implies a clear disadvantage for poor countries. This illustrates the argument put forward above, that poor countries tend to have lower energy effectiveness than rich ones. Actually, the figure shows that there is a substantial difference between regions in this matter. We also note that as a consequence of these differences in effectiveness, especially poor regions are better off when applying the vertical principle than the egalitarian principle.

The Kyoto Protocol from late 1997 represents the first step in an attempt to practise principles of fairness internationally when distributing commitments. To what extent are the principles discussed above applied in this protocol? At first glance, it is difficult see the principles on which the protocol is based. The protocol implies an average 6 to 8 per cent reduction in the emissions of greenhouse gases in the period of 2008 to 2012, compared to the emissions in 1990 for most of the so-called Annex countries, which include most OECD countries and the former Eastern European countries.

However, it is possible to sort out important elements that can be traced back to the principles of fairness discussed in this chapter. The sovereignty principle is clearly represented by the fact that all commitments take the emissions in 1990 as their point of departure. The separation between Annex and non-Annex countries is a clear expression for the vertical principle. So far, only the richest countries, with some exceptions, are committed to emission reductions. Furthermore, the less wealthy of the Annex countries, the former Eastern Europe, are generally provided very moderate commitments since they have already reduced their emissions sufficiently since 1990. Taking 1997 as a base year, the countries are actually allowed to increase their emissions.

The flexible mechanisms included in the protocol contribute to level out costs, in accordance with the horizontal principle. A more explicit expression of the horizontal principle are the exceptions from the 6 to 8 per cent reductions that most of the countries are committed to. The three countries with highest expected abatement costs, Australia, Iceland and Norway, are allowed to increase their emissions compared with the emissions in 1990. Another interesting exception is that of the Ukraine and Russia, which have committed themselves to stabilize their emissions at the 1990 level. Compared with the expected development in these countries, this hardly requires any measures to restrict the emissions of greenhouse gases at all before 2010. One may speculate that this very moderate target is due to the fact that Russia and the Ukraine have suggested that they may actually gain from a higher average temperature. In that case, the protocol also reflects the horizontal principle applied on the net cost.

Of course, a major element of distributing commitments relates to the negotiating power of the different parties. As pointed out by Rayner et al (1999), however, a point of departure for the rich countries' concern for coordinating climate policy was that poor countries were significantly more vulnerable than rich ones. This may explain why it is relatively easy to relate important issues in the negotiations to principles of fairness. As we have seen, different principles of fairness are being used, and will be used in the future, to assess the distribution of commitments to reduce emissions of greenhouse gases. However, whether such a mixture strengthens or weakens the overall fairness of the policy is certainly not appropriately addressed by an economic approach.

Concluding Remarks

Economic analyses of resource allocation seldom address questions of fairness embedded in the initial distribution of the resources. In most cases, economists refer to the existing, or observed, distribution of income, and claim that issues of fairness have to be addressed by others. In the case of establishing a co-ordinated climate policy across countries, such an excuse is hard to accept. A major part of the whole issue is how to distribute the burden of a climate treaty, and the positions taken on other issues raised in the negotiations are closely related to different countries' views on how the commitments should be distributed.

This may lead one to think that economic analysis is inappropriate for evaluating alternative proposals for burden-sharing. This is, however, a too hasty conclusion. Many important aspects of distribution can be addressed properly by the traditional economic analysis of resource allocation. Moral philosophy provides a justification for comparing the burden of different countries, given the existing distribution of income or wealth. In the case of climate policy, this applies to countries on approximately the same level of development. Under this condition, economic analyses have shown how to improve the fairness of a given distribution of emission allowances, in particular by showing the potential of flexible mechanisms.

Economic analysis may also contribute further to the understanding of positions taken by different countries with different expectations, for instance about damage costs. Until now, most studies of optimal climate policy assume full cooperation between all countries of the world. Analyses of the diverging interests among nations will provide a better basis for evaluating burden-sharing, and from an ethical point of view as well.

The greatest challenge for economic analysis of climate policy seems to be to integrate moral philosophy and theories of social choice. This problem has a long history in economics and is of crucial importance for analyses of public policy within countries. When applied to a country, however, one may defend the use of a welfare function by assuming that it reflects the priorities of some public body, for instance a democratically chosen government. There is no such body for the world. A coordinated policy is a result of a negotiated agreement based on voluntary participation. A better representation of alternative ethical principles in economic analysis cannot, therefore, 'solve' the problem of burden-sharing, but may contribute to a more successful distribution of commitments across countries.

References

Aaheim, H A, 1997, *Many Gases and Many Measures: Choice of Targets and Selection of Measures in Climate Policy*, CICERO Report 1997:4, Oslo

Aaheim, H A, Aunan, K and Seip, H M, 1997, Social benefits of energy conservation in Hungary: An examination of alternative methods for evaluation, CICERO Working Paper 1997:10, Oslo

Alfsen, K H, Brendemoen, A and Glomsrød, S, 1992, Benefits of climate policies: Some tentative calculations, Discussion Paper No 69, Statistics Norway

Arrow, K J, 1951, *Social Choice and Individual Values*, Wiley, New York

Asheim, G B, 1994, Net national product as an indicator for sustainability, *Scandinavian Journal of Economics* 96:255–258

Asheim, G B, 1993, Sustainability: ethical foundations and economic properties, mimeo, Norwegian School of Economics and Business Administration, Bergen

Brekke, K A, 1994, Net national product as a welfare indicator, *Scandinavian Journal of Economics* 96:241–252

Bureau of Industry Economics, 1995, *Greenhouse Gas Abatement and Burden Sharing – An analysis of efficiency and equity issues for Australia*, Research Report No 66, Australian Government Publishing Service, Canberra

Burtraw, D and Toman, M A, 1992, Equity and international agreements of CO_2 containment, *Journal of Energy Engineering* 118(2) August

Cline, W R, 1992, *The Economics of Global Warming*, Institute for International Economics, Washington, DC

DFAT and ABARE, 1995, *Global Climate Change – economic dimensions for a cooperative international policy response beyond 2000*, Department of Foreign Affairs and Trade and the Australian Bureau of Agricultural and Resource Economics, Canberra

Ekins, P, 1995, Rethinking the costs related to global warming: A survey of the issues, *Environmental and Resource Economics* 6(3):231–277

Fankhauser, S, 1995, *Valuing Climate Change: The Economics of the Greenhouse*, Earthscan, London

Fankhauser, S, Tol, R S J and Pearce, D W, 1997, The aggregation of climate change damages: A welfare theoretic approach, *Environmental and Resource Economics* 10:249–266

Helm, C, 1999, Applying fairness criteria to the allocation of climate protection burdens: An economic perspective (Chapter 5 of this book)

IPCC (Intergovernmental Panel on Climate Change), 1996, *Climate Change 1995: Economic and Social Dimensions of Climate Change*, Contribution of Working Group III to the Second Assessment Report of the IPCC, Cambridge University Press, Cambridge

Kram,T and Hill, D, 1996, A multinational model for CO_2 reduction, *Energy Policy* 24(1):39–51

Manne, A S, Mendelsohn, R, and Richels, R, 1994, MERGE: A Model for Evaluating Regional and Global Effects of GHG reduction policies, in: N Nakicienovic, W D Nordhaus, R Richels, and F L Tóth, eds, *Integrative Assessment of Mitigation, Impacts, and Adaptation to Climate Change*, CP-94-9 IIASA, Laxenburg

Nordhaus, W D, 1991, To slow or not to slow: The economics of the greenhouse effect, *Economic Journal* 101(407):920–937

Nordhaus, W D, 1993, Rolling the DICE: An optimal transition path for controlling greenhouse gases, *Resource and Energy Economics* 15(1):27–50

OECD/IEA, 1994, *World Energy Outlook*, 1994 edition

Rawls, J, 1971, *A Theory of Justice*, Harvard University Press, Cambridge, MA

Rayner, S, Malone, E and Thompson, M, 1999, Equity issues and integrated assessment (Chapter 2 of this book)

Rose, A, 1992, Equity considerations of tradable carbon emissions entitlements, in: UNCTAD, *Combating Global Warming*, Geneva

Sen, A K, 1979, Personal utilities and public judgments: Or what's wrong with welfare economics?, *The Economic Journal* 89:537–558

Tol, R S J, 1993, The climate fund – Survey on literature on costs and benefits, Working Document W93/01, Institute for Environmental Studies, Vrije Universiteit, Amsterdam

Titus, J, 1992, The cost of climate change to the United States, in: S K Majumdar, L S Kalkstein, B Yarnal, E W Miller, and L M Rosenfeld, eds, *Global Climate Change: Implications, Challenges and Mitigation Measures*, Pennsylvania Academy of Science, Penn

Torvanger, A, Berntsen, T, Fuglestvedt, J S, Holtsmark, B, Ringius, L and Aaheim, H A, 1996, *Exploring Distribution of Commitments – A Follow-up to the Berlin Mandate*, CICERO Report 1996:3, Oslo

Weitzman, M L, 1976, On the welfare significance of national product in a dynamic economy, *Quarterly Journal of Economics* **90**:156–162

7 BIASES IN ALLOCATING OBLIGATIONS FOR CLIMATE PROTECTION: IMPLICATIONS FROM SOCIAL JUDGEMENT RESEARCH IN PSYCHOLOGY

Volker Linneweber

'FAIRNESS' AS OPERATIONALIZATION OF JUSTICE PRINCIPLES

Focusing intrapersonal, interpersonal and intergroup processes, research on fairness-related aspects is well established in social psychology. Studies concern: distributive or procedural justice, reciprocity in altruism, negative reciprocity in interpersonal and intergroup conflicts, and equity or equality. It is thus not surprising that psychology – or, to be more exact, social psychology – is considered relevant for describing and explaining distributions of environment-related benefits and/or burdens in social systems, too.

This chapter briefly outlines levels of analysis and fairness-related justice concepts, before it discusses consequences of applying these to the topic in question and outlines major contributions of psychology to explain social systems' functioning under global change.

Level I of analysis: the individual

Comprehensive research focuses individuals' concepts of justice, their development in the ontogenesis within a broader developmental frame of moral judgement and reasoning (Sigelman and Waitzman, 1991), as well as intra- and interpersonal variance in judgemental preferences. Studies on the individual level explore judgemental procedures and heuristics, and conditions and consequences of perceptions of injustice and unfairness. The theoretical background for this research usually forms existing knowledge on person perception and attribution; its focus is on intra-individual variance (in the ontogenesis or across situations) or on inter-individual differences.

Fairness judgements thus have to be conceptualized as a type of social information-processing sequence, including descriptive, inferential, as well as evaluative components. Corresponding with higher levels of moral development and/or judgement, individuals increasingly utilize comprehensive amounts of information, and they shift from more effect-related aspects (such as benefits or damages) to cause-related aspects (attributes of causing events and/or agents). Higher levels of judgement additionally include attributes of those receiving benefits or those suffering under burdens. Also, the impact of justice-relevant events on the social/societal system in question, as well as ethical principles, are increasingly incorporated in judgements on higher levels (Kohlberg and Diessner, 1991).

Within the last two decades, emphasis has been placed on errors and biases in social judgement. Research shows that social information processing is incomplete, selective, stereotypical, and suffers from motivational as well as cognitive sources for mistakes (Kruglanski and Ajzen, 1983). Dörner (1995; Dörner et al, 1983) shows that different individuals make similar errors when they necessarily reduce complexity. Failure is not random but systematic. This definitely applies to judgement under uncertainty, as Amos Tversky and colleagues impressively have shown (Tversky and Fox, 1995; Tversky and Kahneman, 1974; 1978; 1983). Since the topic in question here is substantially defined by ambiguity, and since this chapter utilizes results from research of judgement under uncertainty as basic building blocks, the author will return to this later in more detail.

Level II of analysis: dyads and small groups

Fairness judgements definitely concern N>1 entities. Hence, social psychology consequently introduced a further level of analysis by studying dyads – primarily close relationships (Lerner and Mikula, 1994; Mikula, 1990; 1992; Mikula and Heimgartner, 1990), but also larger social units in natural contexts or artificial configurations in laboratory settings. At this second level of analysis, individual-centred concepts are applied to dyads. Social judgement is not merely the insulated act of involved actors or outstanding observers. Rather, social judgement significantly contributes to the dynamic of social systems – it fuels the course of interactions. Involved actors mutually evaluate relative input–output ratios and utilize these for justifying own entitlements or – in the case of own relative privilege – realizing obligations.

Research on altruism (Montada and Bierhoff, 1991), as well as on aggression and conflicts (Mummendey, 1984), shows the importance of these processes: altruistic behaviour and the willingness to donate is likely to follow perceptions of having an advantage (compensation) (Bierhoff and Montada, 1988), while instigating a conflict or being aggressive is evaluated as more legitimate in case the actor is disadvantaged (Mummendey et al, 1984b; Mummendey and Otten, 1989). Thus actors necessarily are interested in (re)presenting themselves at a disadvantage in order to legitimize prevailing against each other.[1] Fairness evaluations emphasizing one's own disadvantage are well established means for this (Thompson and Loewenstein, 1992).

---

1 The typesetting of '(re)present' indicates that actors 'present' themselves to others as well as perceive themselves ('represent') in a certain way.

Justice-based evaluations thus have important consequences for micro-social systems (Mikula, 1990; Mikula, 1992; Mikula and Heimgartner, 1990). This does not only apply to isolated judgemental processes but to the dynamic of social systems' functioning in terms of mutual influence processes, use of common resources and allocating shares of negative impacts or protecting them.

Level III of analysis: organizations

In the context of mesosocial systems such as institutions and organizations, a variety of processes imply aspects of distributive or procedural justice (Cropanzano, 1993; Liebig, 1997; Tyler, 1991; Vermunt and Steensma, 1991). Personal selection (Gilliland, 1993), allocations of payoffs (Headey, 1991; Moore, 1991), profit-sharing, promotion prospects, distribution of spheres of influence, structure-building decisions (Daly, 1995) during the life cycle of settings (Wicker and King, 1988), as well as the distribution of power and conduct of conflict (Kabanoff, 1991) are essentially based upon justice principles, including equity, equality and fairness evaluations. Sias and Jablin (1995) conclude that fairness perceptions are often socially constructed by work group members through discourse. Besides aspects within organizations, specific functions of institutions, such as educative or corrective tasks, have been studied extensively. Decisions of superordinate persons are evaluated in terms of fairness. Also here, fairness evaluations and conceptions of procedural as well as distributive justice result from comprehensive lay epistemic processes (Baron, 1990; Goldman, 1986; Gould, 1983; Higgins, 1990; Kruglanski, 1980; 1989; 1990a; 1990b; Kruglanski and Klar, 1987), which are likely to be biases motivationally and cognitively (Kruglanski and Ajzen, 1983).

Level IV of analysis: intergroup relations

Besides the levels discussed above, social comparisons have more recently been studied in intergroup relations (Azzi, 1992). The social identity theory (SIT), introduced by Tajfel (Tajfel, 1982; Tajfel and Fraser, 1978) and Turner and Giles (1981), has been highly influential for European social psychologists within the last decades. The study of intergroup relations is not restricted to interacting groups but includes interacting individuals of different groups, too. One of the core assumptions of the SIT states that depending upon situational components concerning group attributes (such as competition or simply comparative accounts), group membership becomes 'salient' and individuals act (judge, decide) on the basis of their membership. 'Ingroup favouritism' and 'outgroup discrimination' have been identified as stable biases in intergroup relations. For the topic in question, this theoretical background is highly important in various cases: judgements of fairness in defining obligations (for example, means of climate protection) and entitlements (emission rights) are necessarily based on social categories (industrialized countries versus developing countries; Europe versus US; France versus Germany; citizens of towns versus rural areas, etc). Differentiations of aggregated actors are necessary for understanding the system's functioning and for identifying potentials for environmental protection. At the same time, however, the individuals who negotiate are members of social categories themselves and of course intend to represent their interests. Environmental–economic accounting – as a basis for fairness ratings – has thus to be considered as systematically biased.

Justice principles affecting fairness evaluations

Besides the fact that – irrespective of the level of analysis – fairness judgements of competing individuals or groups necessarily are biased egocentrically, three different justice principles (equity, equality and need) have been identified (Wagstaff, 1994). Results are equivocal and highly dependent upon the topic in question, the research procedure, and – probably – are culture specific, which is relevant for the given topic as soon as cross-national or cross-culture accounts are aspired to (Leung et al, 1992). Marin (1985:593) concludes:

> ...all ... [subjects] ... preferred an equitable over an egalitarian allocator. An equitable allocation was also perceived as fairer than one based on equality. When [subjects] were asked to distribute rewards, they chose with greater frequency the equity norm. Although Indonesians and Americans did not differ in their overall preference for equity over equality, Indonesians reported the need to consider the recipient's need, effort, and luck to a greater extent than Americans.

Kashima et al (1988), however, found that in both the cultures they studied (Japan and Australia), subjects were universalistic rather than relativistic in their judgements. Sondak et al (1995) report equity being more commonly used to allocate burdens than benefits. Generally – and definitely with respect to the topic given here – further research is needed, as this chapter discusses below. In accordance with Vermunt and Steensma (1991), this author preliminarily concludes that justice definitely has societal as well as psychological origins, and these have to be considered in our context.

Topic-related research on social justice: the commons dilemma

In social sciences literature on green justice (Opotow and Clayton, 1994), the expressions 'tragedy of the commons' (Hardin, 1968; Tsai, 1993) and 'social trap' (Platt, 1973) have been introduced into the discussion on 'global commons' (Rayner, 1991) – in particular with reference to fishing (Ernst and Spada, 1993; Spada and Opwis, 1985a; 1985b) and the use of water (Thompson and Stoutemyer, 1991). Research has investigated what instigates people or groups not to aim for short hedonistic tendencies (Edney, 1980; Grzelak, 1994; Low and Heinen, 1993; Martichuski and Bell, 1991; Mosler, 1993; Stern, 1978b; Wiener, 1993). Rasinsksi et al (1994) cross-culturally analysed the public support for government spending on the environment. Unfortunately, these studies focus on conflicts of individual versus group interests after having stated that the resource is exploitable but limited. Only some authors study dyads, such as Knapp (1994). He shows that a resource survived best under conditions of cooperative comments, conservative and reciprocally imitating exploitation.

Psychological aspects of fairness-related evaluations that stress common resources have been discussed on the local (Elster, 1992), but not yet on the global, level. Facing the specific quality of the latter – primarily its extremely extended time horizon (Tóth, 1997) – a systematic, theoretical analysis on the use of global commons from a psychologist's perspective, however, is overdue. It must integrate several interconnected attributes:

- the physical quality of the global common 'atmosphere', including its time-related dynamic;
- its perceptive and evaluative quality;
- properties of its use: capability to absorb greenhouse gas;
- properties of its misuses: global warming with various effects in different regions;
- its quality as a 'common', ie causation of change by the activities of different human groups or social categories, with dissimilar vulnerabilities to impacts and different abilities to pay for abatement and thus climate protection.

These components and their interrelations have to be discussed. In the given context, the goal is contributing to an understanding of negotiations about the allocation of obligations and entitlements, and burdens and benefits in using and protecting global environmental resources.

As increasingly confirmed, psychology accepts the challenge and applies state-of-the-art concepts in order to contribute to global change research (Boniecki, 1977; Busemeyer and Myung, 1987; Fischhoff and Furby, 1983; Kruse, 1995; Pawlik, 1991; Sjöberg, 1989; Sloan, 1992; Stern, 1978a; 1978b; 1992; Stokols, 1992; Vining, 1987). It is accepted that applying psychological insight should not be limited to what may be termed classic environmental–psychological areas, such as immediate daily concerns and environmental awareness, but should be extended to include global relationships between man and the environment. In light of an increasingly interdependent world population, further potential exists to extend the area of social–psychological justice and conflict research beyond its existing basis of analysis of interpersonal and intergroup relationships, and to test further possibilities for application on the level of larger units (nations, cultures, generations).

'ENVIRONMENT' AS A CONCEPT

Conceptualizing the atmosphere as a global common, and considering 'fair' proportions of its utilization and protection, it should be stressed that the components are not a priori defined, but rather scientifically, culturally and socially constructed. Their definition and use includes considerable potential for differences of interpretation that are relevant for fairness-related judgement. Since impacts on the atmosphere are perceived as one facet of the man–environment interaction, it is essential to analyse *everyday concepts of environment* on different unit levels (from individuals to groups and from societies to cultures) in order to understand fairness-based negotiations.

Environment is not an absolute but a relative concept. Its definition is not restricted to the common in question, but based on the potential for social system's functioning (for example, in the production of foodstuffs and energy) and environmental impacts on social systems. Environment is thus jointly conceptualized with its potential for utilization as well as with its fragility. Various studies have shown that analyses of such *relative concepts of environment* are adequate for investigating man–environment relationships on different unit levels, since 'if men define situations as real, they are real in their consequences' (Thomas and Thomas, 1928:572).

The historical development of the concept 'environment'

That people direct their actions with regard to their effect on what they define as 'the environment' is – historically speaking – new and even presently not at all self-evident. While introduced to characterize irritating developments within immediate surroundings, the focus on regional conditions and changes soon proved to be too narrow. Particularly in connection with the deterioration of water quality in coastal areas and with acid rain, it increasingly became clear that cause and effect in critical environmental changes were not necessarily confined to the same place, but could have supraregional importance. Burdening of coastal waters is connected with substances carried great distances by rivers, originating from the use of fertilizers and pest control substances in agriculture, or from other contamination of surface waters (Bossel, 1990:135). Acid rain in Scandinavia may largely be traced back to emissions of SO_2 and NO_x in western Europe; similar phenomena in Canada have their origins in the eastern US. Problems in Alaska result from emissions in Japan.

That man-made occurrences may be global in scale became apparent to the general public, in particular in connection with the greenhouse effect and ozone depletion. Since the Rio conference at the latest, it has become clear that the essential environmental problems we are confronted with may only be adequately comprehended in their global dimensions. This is definitely true of global dynamics of the atmosphere. *Ab initio,* local definitions of problems hence became regional, then interregional and finally global. For our argument, it is important that dimensions are not objectively set but result from social and societal constructions of problems.

Temporal as well as geographic scales exploded: global circulation models (GCMs) indicated that long-term calculations must be introduced when global processes are under consideration. Environmental economists (Broome, 1992; Cline, 1992; Manne and Richels, 1992; Tóth, 1995) consistently underlined the necessity of including long-term effects in models calculating the cost of global warming. Deliberations on 'reasonable' use of global resources also endeavoured to anticipate the interests of future-user generations and, in considering, for instance, the discounting of investments towards the compensation or reduction of environmental burdens, posed considerations on attaining intergenerative justice (Lind, 1995; Schelling, 1995; Tóth, 1994). Both scientists and politicians are thus confronted with a problem which we will consider in the following sections: the further prognoses reach into the future, the more they are bound together with uncertainties. Synergetic effects, in particular, may only be modelled with difficulty, and it is scarcely possible to predict the ability of human users to adapt.

Ambiguity of interpretation

With regard to commons-related negotiations and conflicts about obligations and entitlements, it will be shown that the problems are not to be reduced due to a lack of scientifically precise prognosis. Rather, conditions attended by a high level of uncertainty directly influence altercations concerning emission rights as well as reasonableness of climate protection investments; calculations and decisions follow rules that are of social origin and of social significance. The ambiguity of interpretation arising from uncertainty is systematically exploited in order to increase the favourability of own positions respectively. These tactics or strategies of social comparison have already been identified in various everyday situations, and it has been shown that they determine unfairness evaluations in the course of conflicts. Here, some of these observations will be transferred to the subject under consideration.

USE OF ENVIRONMENT AS A PHENOMENON OF (INTER)DEPENDENCE

Meanwhile it is a self-evident position of global change research that humans are causing, as well as are affected by, global and regional environmental changes (WBGU, 1993). Thus they are 'objectively' mutually dependent in their use of the environment.

Trying to understand debates or operationalizations of fairness among interdependent actors, we have to analyse the conditions, the features and the consequences of the perception of interdependence through the use of environment – or, in other words, *subjective* mutual dependence. While research on the commons dilemma has investigated the features and consequences, the conditions of the cognition of interdependence, on the other hand, have hardly been looked at. Rather, the fact that a common is fragile and that the users mutually influence one another is experimentally induced. When, however, features of the commons dilemma are of interest, one must consider the involved actors' conceptualization of interdependence. Debates or conflicts on the respective utilization of resources become more transparent if biases in constructions on interdependence have been identified (Linneweber, 1995).

Self-monitoring ability of human users of the environment

A feature of man's distinct place in nature (Scheler, 1926; 1928) is his ability to recognize having been influenced by the environment in the past, being influenced in the future and influencing the environment in turn. It is tempting to classify this competence as cognitive and restricting it to single individuals.

This limitation is not intended. Discussing use of environment, influences and cognitions, judgements, assessments, and justifications, refers both to individual users and their aggregates (groups, nations, cultures, generations). Results of (self-) reflection processes are, of course, expressed assessments, attributions of blame, etc. These are essentially determined by supra-individual features (such as social consensus assessments) social, societal, cultural and economic factors.

Self-monitoring as a product of an 'active filter'

The basic capacity for self-reflection should not be confused with the capability to do this 'correctly' (guided by the status of current scientific research, for example) or moreover with the ability to recognize the limits of the potential for self-monitoring (Luhmann, 1986; Scheler, 1926). Rather, biases are likely. This chapter will discuss how systematic mistakes and errors may arise and what functional significance these have for defining obligations and entitlements. Illustratively, one can speak of an 'active filter' which 'produces' information on environment use. Active means that information on man–environment interaction may as well be added as omitted, and may be accentuated and also deemphasized.

In industrial nations, but also increasingly in countries in transition and in developing countries, environmental problems are communicated primarily through the media. Media-internal and media-specific processes are of high significance (Bell, 1989; 1994; Beniger, 1993; Diggelmann, 1986; Diggelmann and Domont, 1986; Diggelmann and Schanne, 1988; Dumas and Gendron, 1991; Kempton, 1991; Oodit and Simonis, 1993; Schanne and Meier, 1992; Singer and Endreny, 1993) and are interlinked with social-, cognitive- and environmental-psychological processes as discussed here.

Conceptualizing the performance of this active filter, the *perspectives* on the environment must be differentiated: humans are – and perceive themselves as being – agents of, and affected by, the dynamics of the Earth system. Models have been developed in the cognitive sciences that are capable of illustrating processes of *perspective-specific* perception and judgement. Applying these to the given context, these explain:

- biased representation of the actual state of the Earth system: motivational and cognitive strategies ignoring undesired functional relationships underassessing their significance;
- biased representation of the probable developments of the system: motivational and cognitive strategies permitting an underestimation of the probability of undesired scenarios occurring and overestimating the reversibility of threatening developments or acute situation.

The implications of position-specific (user versus person affected) divergent perceptions on fairness evaluations will be discussed later. Firstly, the question of which 'paths' are taken by perspective-specific perceptions and judgements will be addressed.

Positive illusions as motivated distortion

The active filter of 'perceptions and judgements of climate change' has a significant function in dealing with threatening information, such as indications of irreversible damage, likely shortages or reevaluations of critical tendencies. Taylor (1989) points out that self-deceiving 'positive illusions' make dealing with undesired information possible. It can be assumed that these relate not only to current conditions, but also to past conditions ('hindsight bias'; Fischhoff, 1975; Hawkins and Hastie, 1990; but also, under certain circumstances, its reversal: Mazursky and Ofir, 1990) and to future developments (Brown et al, 1993). This may be expected particularly in regard to features characterizing the subject discussed here: fuzzy dynamics, uncertainties (Tversky and Kahneman, 1974; 1978) and self-relevant issues (Mark and Mellor, 1991).

Assessing the impact of human activities on the atmosphere and its feedback on human societies, the general tendency may be expected that:

- the probability of negative alterations in the status quo is underestimated; and
- the seriousness of consequences is disputed, if they cannot be ignored.

These tendencies likely occur and are 'successfully' applied, particularly under conditions of informational or evaluative uncertainty (Tversky and Kahneman, 1974; 1978). Regarding the consequences linked to greenhouse gas emissions into the atmosphere, such conditions are present. In negotiations about fair distributions of climate protection efforts, they are systematically used in negotiations and conflicts as possibilities for degrees of freedom for interpretation in order to emphasize the legitimacy of the position of each party in the 'user community'.

The thesis here expounded of positively distorted perception and assessment is based on a motivational assumption. It proceeds mainly in agreement with Taylor and Brown (1988). They argue that positive illusions of the self, the world and the future are essential for mental health. The idea that mankind is heading for an ecological or climatic catastrophe, that mankind will, in the future, be affected by the depletion of stratospheric ozone in the form of illness, or that mankind is faced with a dramatic shortage in essential resources resulting from climate change, is extremely threatening. It creates a motive to ignore critical developments. A less problematic perspective is achieved by:

- denying that threatening developments have actually occurred or will actually occur;
- doubting their effective danger by assuming, for example, that planetary ecosystems (or future users) are able to counter apparently threatening developments (regenerative or compensatory ability), and that human users have the power, or will develop it, to cope with threatening developments.

Further assumptions allowing differentiations will be presented further on, and this chapter will refer to this, discussing the impact of positive illusions on defining 'fair' distribution obligations and entitlements from each point of view.

Limited capacity in processing information as a cause of cognitive distortion

A further shortcoming in human self-reflection may be seen in a limited capacity to process information (Dörner, 1985; Dörner et al, 1983). Differing from the motivational potential to ignore threats discussed above, the concern here is with cognitive aspects. These are particularly significant in relation to complexity, multidimensionality and uncertainty. Studies on lay epistemology show that:

- monocausal explanations are preferred ('reduction of complexity');
- preexisting explanations are preferred rather than new models developed ('availability heuristic'; Tversky and Kahneman, 1974);
- aspects deemed to be representative of a phenomenon are overvalued ('representativeness heuristic'; Nisbett and Ross, 1980).

Cognitive tendencies towards simplification have already been identified in relatively simple causal relationships: the interaction effects of two independent variables on one dependent variable are likely to be transformed into two main effects. It is thus not surprising at all that complex interrelations or non-linear developments are prime candidates for simplification. We must assume that such cognitive simplification strategies are significant in the fairness-based discourse (for instance, in judgements of legitimacy or calculations of effect). It is also assumed that simplification (such as over or underestimation of effects, extrapolations of past or future developments) are not coincidental but follow an identifiable system.

Human actors constructing (inter)dependence

Originators and subjects of global change are, via the use of global resources, *interdependent*. No absolute, long-term assignment into one category or the other is intended in this definition of position. Individual actors (or aggregates) may equally be originator and victim at the same time and – in temporary or object-based differentiation – either originator or victim. While extreme events such as the Chernobyl accident or acid rain allow interdependence structures to be seen clearly, this is not the case for global (and thus vague) developments such as the depletion of stratospheric ozone, global warming or the loss of biodiversity. Here, (inter)dependence relations result instead from constructions which likely are systematically distorted. The interdependence relationship between originators and persons subject to critical developments of the Earth system forms the basis for use-based legitimacy judgements, such as the formulation of claims and expectations of preventive or compensatory measures. In order to explain the positions taken in negotiations about fair distributions of obligations and entitlements, which are drawn essentially from such legitimacy concepts and balancing, it is essential to analyse 'constructions of interdependence'.

Psychological Dimensions of Global Climate Change

Continuing Pawlik's (1991) analysis, this chapter relates the processes previously discussed to the particular 'quality' of global change – or at least to those parts of it potentially resulting from human use. It has already been indicated that judgements concerning effects, identification of responsible persons/groups, forecasts and considerations of legitimacy must be viewed as resulting from systematically distorted and simplified 'naive' models (Greenwald and Banaji, 1995; Schlenker et al, 1994). The following section deals with those characteristics of global change which form or which permit the formation of the resulting 'naive models'.

Natural sciences and social vagueness

Uncertainties in natural scientific analyses have already been mentioned. Global circulation models (GCMs) still contain ambiguities – for example, concerning the role of the oceans. Only recently, an unidentified CO_2 sink has been identified: the biomass on the African continent had been incorrectly calculated according to forestry criteria. These areas possibly contain double the amount of biomass previously supposed and therefore have an appreciable greater CO_2 absorption capacity than had previously been assumed in GCMs (Pearce, 1994). Global environmental changes are in any case imperceptible to individual actors, and even scientifically difficult to establish. It is still controversial what portion is anthropogenic ('low signal-to-noise-ratio'). The scope this ambiguity creates for naive models has already been discussed; consequences for position-taking in negotiating fairness will be examined later.

Global environmental change is additionally fuzzy in a *social* sense (Wit, 1994). The identification of relevant influences and accountability usually is difficult. Even in the case of local occurrences such as water pollution, responsible agents may be difficult to identify. Even more demanding is the identification of accountable involvement in global developments or relative involvement in cumulative effects, direct or indirect consequences and synergetic processes. Social uncertainty exists not only on the side of those causing, but also on the side of those affected by environmental change. Studies on legitimacy and justice in connection with exposure to local environmental changes or the regional effects of global environmental changes ('environmental equity', Anderton et al, 1994a; Anderton et al, 1994b; Kasperson and Dow, 1991; Zimmerman, 1993) reach highly complex conclusions. They definitely contradict the attempt to categorize 'clearly disadvantaged' versus 'clearly privileged'. An even more apparent lack of clarity exists with regard to the 'level of affectedness' by global changes. This is also apparent in balancing fair burdens and benefits for climate protection based on environmental–economic accounts.

Perceptive temporal and spatial indirectness

Attribution research indicates that humans have the propensity to examine phenomena for their visible *causes*. Although the universal application of this motive is sometimes doubted (Bond, 1988; Hamilton and Hagiwara, 1992), the search for causes and effects with the aim of achieving an *insight into their interconnections* characterizes the worldview of highly developed civilizations. The cognition of causal relationships is not an aim in itself, but a precondition for the potential to control and influence developments. Through exerting influence on recognized causes, undesired phenomena may be prevented or desired developments promoted. Indirectness and multidetermination and a wide divergence in spatial or temporal circumstances complicate causal attributions (Schlenker et al, 1994). Indirectness influences the attribution of responsibility; it my be denied where it de facto exists (β-error), or may be attributed where it de facto is absent (α-error). Changes immediately perceptible, and those with visible and/or clear cause–effect relationships, are more potent behavioural incentives than those involved in 'constructed' cases. Concerning debates on fair burden-sharing, it should be obvious that individuals, groups or states are more likely to engage themselves if they experience immediate impacts (such as small island states). They are, moreover, interested in attributing responsibility to actors who have benefited in the past (industrialized nations). Furthermore, they claim that industrial nations are highly responsible for critical states or trends of the atmosphere and hence postulate high engagements in climate protection as a compensation.

Scarcity of clearly indicative occurrences

Climate change effects only rarely become immediately obvious. Global warming has regional effects; but in the case of 'spectacular' events such as droughts, floods and loss of biodiversity, global warming competes as a causal explanation with regional features and/or coincidentally unfavourable circumstances. Only exceptional directly apparent phenomena may *convincingly* and *clearly* be evaluated as the consequences of anthropogenic impacts on the atmosphere. Studies on 'representative heuristic effects' (see above) point out that this property of climate change phenomena makes classifying global trends as causes difficult, if not impossible. Global changes are more likely to be characterized by lack of certainty, scientific controversy and long-term trends (which possibly will not affect the current generation) than by memorable 'representative' events. This, of course, implies that concerning fair distributions of climate protection burdens, negotiations start with defining the problem and its assessment. Compared with other environment-related topics with unequivocal cause–effect relations and immediate, as well as local, effects, this 'masking' opens significant degrees of freedom for biases in debating the issue.

Distance between actors and those affected

In terms of global scale there is both a temporal and a spatial distance between actors and those affected. Current model calculations extend far into the next century and beyond. Even if a drastic reduction of CO_2 emissions were to occur – in particular in the industrial nations – at the present time, an observable effect could be expected only within decades, as a result of the inertia of the Earth system. Endeavours to reduce CFC emissions will also have only middle-term effects. This temporal distance implies that actors presently responsible will not experience direct confirmation of the effects of endeavours to reduce emissions, which otherwise might have functioned as a stimulus for further efforts. On the contrary, the effects of 'learned helplessness' (Seligman, 1975) must be expected: the non-appearance of direct confirmation of endeavours in spite of the sacrifice involved may lead to a decreased inclination to continue them. This is at least true on the individual level. Perhaps those deciding on the basis of scientifically based scenarios will introduce more sophisticated and future-oriented considerations.

The *spatial* distance between actors and those affected has implications not only for denying responsibility. The effects are also indirectly *perceived*. It is no coincidence that interest groups such as the small island states are profiled by claiming effective means against, or investment as compensation for, past damage from the industrialized nations. If negative effects of misusing the atmosphere were more direct, a different risk assessment, as well as higher acceptance for introducing preventative measures, might be expected.

Weak stimulus of non-direct egocentric behaviour

A significant reduction of greenhouse gas emissions is also unattractive since – as typically in the commons dilemma – a direct 'reward' is missing. Spectacular altruistic actions such as direct aid in crisis situations, or impressive aid programmes, are prestigious for their instigators and may enhance the self-esteem of groups involved. Thus they are positive with respect to social comparisons even if they include immediate benefits for those conducting the activity. Non-egocentric decisions and modes of behaviour towards the atmosphere, however, in the first instance imply apparent disadvantages, particularly economic ones (see the debate about high-cost versus low-cost decisions: Kushler, 1983).

Not without good reason is it stressed by the environmentalists in the debate on economy–ecology conflicts that – at least in the long term – no alternative exists. It is pointed out that the development of environmental protection technology will, in the long term, bring economic advantages, in addition to the positive effects resulting from their more thorough application on the spot. It is clear from these discussions that a stimulus is not directly perceptible, but has to be perceived through a 'concept' – commons dilemmas are essentially characterized by indirect cause–effect relationships of this kind.

IMPLICATIONS OF DIVERGENT PERSPECTIVES IN JUSTICE-BASED FAIRNESS EVALUATIONS

In man–environment relationships, people assume either the role of actors or of those affected (WBGU, 1994). Assignment into one group or the other is not permanent, but varies depending upon circumstances. In connection with worldwide greenhouse gas emissions, 'the industrial nations' are to be seen as actors while developing nations primarily are affected (Bossel, 1990). With regard to population and economic growth and its associated problems, such as increasing CO_2 emissions, the position is reversed – at least when in future-oriented scenarios. In discussing intergenerational justice, the position of the actors and those affected are at least logically unequivocal. The positions, however, collapse in regard to problems which also affect those who cause them. And while this is immediately evident in connection with the regional impact upon the environment, the simultaneousness of the positions of actor and affected is fuzzy both on a global and long-term scale in light of using the atmosphere. With increasing *dimensionality* of interrelations, increasing *vagueness* of cause–effect relationships and increasing *lack of certainty* regarding expected trends, this creates the potential for the *accentuation* of the positions of either actor or those affected. This is highly important for evaluations of fairness.

Divergence of perspective

The assumption of a divergence in perspective between actors and those affected is of central importance in explaining the dynamics between individuals or social units (Mikula, 1994; Mummendey et al, 1984a; Mummendey and Otten, 1989). Divergence of perspective signifies that circumstances (impact on the atmosphere and disadvantages by allocating obligations for climate protection) are differently perceived and assessed by actors (A) and those affected (B). A systematic difference of judgement may be expected, in particular, with regard to the effects produced by A on B. In interpersonal relations this has been proved, both in immediate interaction as well as in interdependence relationships configured by using the environment (Linneweber, 1988a; 1988b).

This chapter assumes that divergence of perspective does not merely characterize incompatible representations of social units involved, but also has consequences for decision-making and policy. In conflicts from the interpersonal to the international level, the parties' own actions are justified as being fair (defensive, retaliatory or even preventative) with regard to the reciprocity norm (Gould, 1983; Gouldner, 1960; Marsh et al, 1978; Rule and Nesdale, 1976), while those of the other party are condemned. The significance of divergent perspectives as an explanation of the dynamic force in social systems becomes apparent.

Perspective of social units affected by a negative event or development

Affected social units are concerned to (re)present themselves as unjustly treated, disadvantaged, etc. They achieve this by:

- dramatizing the damage caused (or prevented by their own efforts or by favourable circumstances);
- stressing the disproportionateness of the influences working on them;
- identifying persons or groups who caused or were responsible for the damage, in order to justify and pursue claims for compensation.

Perspective of social units causing a negative event or development

Actors, on the other hand, try to:

- understate the importance of events or processes by ignoring the damages caused;
- deny having caused the damage; or
- refuse responsibility.

Preference of perspectives

Actors prefer presenting themselves as 'affected'. Determining their position is achieved through segmentation and accentuation of interactive sequences (Mummendey and Otten, 1989; Newtson, 1973; 1976) or by emphasizing that the party's own actions (CO_2 emissions) have solely the function of putting right previous injustices directed against themselves (Bierhoff et al, 1986; Caddick, 1982; Kabanoff, 1991; van Knippenberg and van Oers, 1984; Walster, 1976). Based on the principle of reciprocity, expectations that others have to engage in climate protection can also be justified.

Further alternatives are available in 'multichannel' interdependence relationships. Here it is additionally possible to compare influences of varying quality (such as multiple burdens versus multiple benefits). Not only a quantitative accentuation may be used for a 'self-serving' depiction of the 'own' position. It may also be stressed that the quality of the 'channel' on which the 'own' social unit is underprivileged (and hence in the 'affected' position) is more severe (worse, more damaging, more dangerous, more threatening, but less certain, more long term, of more consequence) compared with the 'channel' on which the social unit exercises (from its point of view) critical influence. The preferred perspective may thus be induced in multidimensional interdependence relationships by correspondingly stressing the importance of judgement dimensions.

It should be obvious that judgemental biases have to be introduced when explaining negotiations of fair partitions of climate protection burdens and emission rights. Uncertainties concerning the system's dynamic, vague and multicriterial topics provide evaluative degrees of freedom perfectly utilized by actors in their constructions of fairness.

REFERENCES AND FURTHER READING

Anderton, D L, Anderson, A B, Oakes, J M and Fraser, M R, 1994a, Environmental equity: the demographics of dumping, *Demography* **31**:229–248

Anderton, D L, Anderson, A B, Rossi, P H, Oakes, J N, Fraser, M R, Weber, E W and Calabrese, E J, 1994b, Hazardous waste facilities – environment equity issues in metropolitan areas, *Evaluation Review* **18**:123–140

Azzi, A E, 1992, Procedural justice and the allocation of power in intergroup relations: Studies in the United States and South Africa, *Personality and Social Psychology Bulletin* **18**(6):736–747

Baron, R M, 1990, Kruglanski's lay epistemic theory: When rules describe but do not regulate, *Psychological Inquiry* **1**(3):200–202

Bell, A, 1989, *Hot news: media reporting and public understanding of the climate change issue in New Zealand: A study in the (mis)communication of science*; Victoria University; Department of Linguistics, Wellington, New Zealand

Bell, A, 1994, Climate of opinion: public and media discourse on the global environment, *Discourse and Society* **5**:33–64

Beniger, J R, 1993, Reporting on risk – how the mass media portray accidents, diseases, disasters, and other hazards, Review by E Singer and P M Endreny, *Communication Research* **20**:870–871

Bierhoff, H W, Cohen, R L and Greenberg, J, eds, 1986, *Justice in Social Relations*, Plenum, New York, NY

Bierhoff, H W and Montada, L, eds, 1988, *Altruismus*, Hogrefe, Göttingen

Bond, M H, eds, 1988, *The cross-cultural challenge to social psychology*, Sage, Newbury Park, CA

Boniecki, G J, 1977, Is man interested in his future? The psychological question of our times, *International Journal of Psychology* **12**:59–64

Bossel, H, 1990, *Umweltwissen: Daten, Fakten, Zusammenhänge*, Springer, Berlin

Broome, J, 1992, *Counting the cost of global warming*, White Horse Press, Cambridge, UK

Brown, R S, Williams, C W and Lees-Haley, P R, 1993, The effect of hindsight bias on fear of future illness, *Environment and Behaviour* **25**:577–585

Busemeyer, J R and Myung, I J, 1987, Resource allocation decision making in an uncertain environment, *Acta Psychologica* **66**:1–19

Caddick, B, 1982, Perceived illegitimacy and intergroup relations, in: H Tajfel, ed, *Social identity and intergroup relations*, Cambridge University Press, Cambridge

Cline, W R, 1992, *The economics of global warming*, Institute for International Economics, Washington, DC

Cropanzano, R, ed, 1993, *Justice in the workplace: Approaching fairness in human resource management*, Lawrence Erlbaum Associates, Inc, Hillsdale, NJ, US

Daly, J P, 1995, Explaining changes to employees: The influence of justification and change outcomes on employees' fairness judgements, *Journal of Applied Behavioural Science* **31**(4):415–428

Diggelmann, A, 1986, *Der SANASILVA-Waldschadenbericht 1986 in den aktuellen schweizer Massenmedien: eine inhaltsanalytische Diffusionsstudie*; Programmleitung SANASILVA, Birmensdorf

Diggelmann, A and Domont, P, 1986, Der SANASILVA-Waldschadenbericht 1986 in den aktuellen schweizer Massenmedien, *Schweizerische Zeitschrift für Forstwesen* **139**:37–52

Diggelmann, A and Schanne, M, 1988, *Der SANASILVA-Waldschadenbericht 1988 in den aktuellen schweizer Massenmedien: eine inhaltsanalytische Diffusionsstudie*; Programmleitung SANASILVA, Zürich

Dörner, D, 1985, Verhalten, Denken und Emotionen, in: L H Eckensberger and E D Lantermann, eds, *Emotion und Reflexivität*, Urban and Schwarzenberg, München

Dörner, D, 1995, Logik des Mißlingens, in: K H Erdmann and H G Kastenholz, eds, *Umwelt- und Naturschutz am Ende des 20 Jahrhunderts: Probleme, Aufgaben und Lösungen*, Springer, Berlin; Heidelberg

Dörner, D, Kreuzig, H W, Reither, F and Stäudel, T, eds, 1983, *Lohhausen Vom Umgang mit Unbestimmbarkeit und Komplexität*, Huber, Bern

Dumas, B and Gendron, C, 1991, Culture écologique; étude exploratoire de la participation de médias québécois à la construction de représentations sociales de problèmes écologiques, *Sociologie et sociétés* **23**:163–180

Edney, J J, 1980, The commons problem: Alternative perspectives, *American Psychologist* **35**(2):131–150

Elster, J, 1992, *Local justice: How institutions allocate scarce goods and necessary burdens*, Russell Sage Foundation, New York, NY

Ernst, A M and Spada, H, 1993, Bis zum bitteren Ende?, in: J Schahn and T Giesinger, eds, *Psychologie für den Umweltschutz*, Beltz, Weinheim

Fischhoff, B, 1975, Hindsight ≠ foresight: The effect of outcome knowledge on judgement under uncertainty, *Journal of Experimental Psychology: Human Perception and Performance* 1:288–299

Fischhoff, B and Furby, L, 1983, Psychological dimensions of climatic change, in: R S Chen, E Boulding and S H Schneider, eds, *Social science research and climate change*, Reidel, Dordrecht

Gilliland, S W, 1993, The perceived fairness of selection systems: An organizational justice perspective, *Academy of Management Review* 18(4):694–734

Goldman, A I, 1986, *Epistemology and cognition*, Harvard University Press, Cambridge, MA

Gould, C C, 1983, Beyond causality in the social sciences: Reciprocity as a model of non-exploitative social relations, in: R S Cohen and M W Wartofsky, eds, *Epistemology, methodology, and the social sciences*, Reidel, Boston, MA

Gouldner, A W, 1960, The norm of reciprocity: A preliminary statement, *American Sociological Review* 25:161–178

Greenwald, A G and Banaji, M R, 1995, Implicit social cognition: attitudes, self-esteem, and stereotypes, *Psychological Review* 102:4–27

Grzelak, J, 1994, *An individual and the commons*, Paper presented at the EAESP small group meeting on social interaction and interdependence; Amsterdam, The Netherlands, April 28–May 1

Hamilton, V L and Hagiwara, S, 1992, Roles, responsibility, and accounts across cultures, *International Journal of Psychology* 27:157–179

Hardin, G J, 1968, The tragedy of the commons, *Science* 162:1243–1248

Hawkins, S A and Hastie, R, 1990, Hindsight: Biased judgements of past events after outcomes are known, *Psychological Bulletin* 107:311–327

Headey, B, 1991, Distributive justice and occupational incomes: Perceptions of justice determine perceptions of fact, *British Journal of Sociology* 42(4):581–596

Higgins, E T, 1990, Lay epistemic theory and the relation between motivation and cognition, *Psychological Inquiry* 1(3):209–210

Kabanoff, B, 1991, Equity, equality, power, and conflict, *Academy of Mangement Review* 16:416–441

Kashima, Y, Siegal, M, Tanaka, K, and Isaka, H, 1988, Universalism in lay conceptions of distributive justice: A cross-cultural examination, *International Journal of Psychology* 23:51–64

Kasperson, R E and Dow, K M, 1991, Developmental and geographical equity in global environmental change: a framework for analysis, *Evaluation Review* 15:149–171

Kempton, W, 1991, Public understanding of global warming, *Society and Natural Resources* 4:331–345

Knapp, A, 1994, *Der Umgang mit knappen Ressourcen*, Vol 12, Hogrefe, Göttingen

Kohlberg, L and Diessner, R, 1991, A cognitive-developmental approach to moral attachment, in: W M K Jacob, L Gewirtz, eds, *Intersections with attachment*, Lawrence Erlbaum Associates, Inc, Hillsdale, NJ, US

Kruglanski, A W, 1980, Lay epistemo-logic – process and contents: another look at attribution theory, *Psychological Review* 87:70–87

Kruglanski, A W, 1989, *Lay epistemics and human knowledge: Cognitive and motivational bases*, Plenum Press, New York, NY

Kruglanski, A W, 1990a, Lay epistemic theory in social-cognitive psychology, *Psychological Inquiry* 1(3):181–197

Kruglanski, A W, 1990b, 'Lay epistemic theory in social-cognitive psychology': Response, *Psychological Inquiry* 1(3):220–230

Kruglanski, A W and Ajzen, I, 1983, Bias and error in human judgement, *European Journal of Social Psychology* 13:1–44

Kruglanski, A W and Klar, Y, 1987, A view from a bridge: Synthesizing the consistency and attribution paradigms from a lay epistemic perspective, *European Journal of Social Psychology* 17:211–241

Kruse, L, 1995, Globale Umweltveränderungen: Eine Herausforderung für die Psychologie, *Psychologische Rundschau* 46:81–92

Kushler, M G, 1983, A response to the BPA evaluation plan: some modest but utilitarian considerations, *Evaluation and Program Planning* 6:115–119

Lerner, M J and Mikula, G, eds, 1994, *Entitlement and the affectional bond: Justice in close relationships*, Plenum Press, New York, NY

Leung, K, Au, Y F, Fernandez-Dols, J M and Iwawaki, S, 1992, Preference for methods of conflict processing in two collectivist cultures, *International Journal of Psychology* 27:195–209

Liebig, S, 1997, *Soziale Gerechtigkeitsforschung und Gerechtigkeit in Unternehmen*, Hampp, München; Mering

Lind, R, 1995, Intergenerational equity, discounting, and the role of cost-benefit analysis in evaluating global climate policy, *Energy Policy* 23(4/5):379–390

Linneweber, V, 1988a, Jeopardizing patterns of settings: deviations and deviation-counterings, in: H van Hoogdalem, N L Prak, T J M van der Voordt and H B R van Wegen, eds, *Looking back to the future*, Delft University Press, Delft

Linneweber, V, 1988b, Norm violations in person x place transactions, in: D Canter, J C Jesuino, L Soczka and G M Stephenson, eds, *Environmental Social Psychology*, Kluwer Academic Publishers, Dordrecht

Linneweber, V, 1995, Evaluating the use of global commons: lessons from research on social judgement, in: A Katama, ed, *Equity and Social Considerations Related to Climate Change*, ICIPE Science Press, Nairobi (Kenya)

Low, B S and Heinen, J T, 1993, Population, resources, and environment: Implications of human behavioural ecology for conservation, *Population and Environment: A Journal of Interdisciplinary Studies* 15:7–41

Luhmann, N, 1986, *Ökologische Kommunikation*, Westdeutscher Verlag, Wiesbaden

Manne, A S and Richels, R, 1992, *Buying greenhouse insurance: The economic costs of carbon dioxide emission limits*, MIT Press, Cambridge and London

Marin, G, 1985, The preference for equity when judging the attractiveness and fairness of an allocator: The role of familiarity and culture, *Journal of Social Psychology* 125:543–549

Mark, M M and Mellor, S, 1991, Effect of self-relevance of an event on hindsight bias: The foreseeability of a layoff, *Journal of Applied Psychology* 76:569–577

Marsh, P, Rosser, E and Harré, R, 1978, *The rules of disorder*, Routledge and Kegan Paul, London

Martichuski, D K and Bell, P A, 1991, Reward, punishment, privatization, and moral suasion in a commons dilemma, *Journal of Applied Social Psychology* 21:1356–1369

Mazursky, D and Ofir, C, 1990, 'I could never have expected it to happen': The reversal of the hindsight bias, *Organizational Behaviour and Human Decision Processes* 46:20–33

Mikula, G, 1990, Austausch und Gerechtigkeit in Freundschaft, Partnerschaft und Ehe: Ein Überblick über den aktuellen Forschungsstand, *Berichte aus dem Institut für Psychologie der Universität Graz* 3:25

Mikula, G, 1992, Austausch und Gerechtigkeit in Freundschaft, Partnerschaft und Ehe: Ein Überblick über den aktuellen Forschungsstand, *Psychologische Rundschau* 43:69–82

Mikula, G, 1994, Perspective-related differences in interpretations of injustice by victims and victimizers: A test with close relationships, in: M J Lerner and G Mikula, eds, *Entitlement and the affectional bond*, Plenum, New York, NY

Mikula, G and Heimgartner, A, 1990, Experience of injustice in intimate relationships, *Berichte aus dem Institut für Psychologie der Universität Graz* 7:26

Montada, L and Bierhoff, H W, eds, 1991, *Altruism in social systems*, Verlag für Psychologie Dr C J Hogrefe, Göttingen

Moore, D, 1991, Entitlement and justice evaluations: Who should get more, and why, *Social Psychology Quarterly* 54(3):208–223

Mosler, H J, 1993, Self-dissemination of environmentally-responsible behaviour – the influence of trust in a commons dilemma game, *Journal of Environmental Psychology* 13:111–123

Mummendey, A, ed, 1984, *Social psychology of aggression: From individual behaviour to social interaction*, Springer, New York, NY

Mummendey, A, Linneweber, V and Löschper, G, 1984a, Actor or victim of aggression: Divergent perspectives – divergent evaluations, *European Journal of Social Psychology* 14:297–311

Mummendey, A, Linneweber, V and Löschper, G, 1984b, Aggression: From act to interaction, in: A Mummendey, ed, *Social psychology of aggression: From individual behaviour to social interaction*, Springer, New York, NY

Mummendey, A and Otten, S, 1989, Perspective-specific differences in the segmentation and evaluation of aggressive interaction sequences, *European Journal of Social Psychology* 19:23–40

Newtson, D, 1973, Attribution and the unit of perception of ongoing behaviour, *Journal of Personality and Social Psychology* 28:28–38

Newtson, D, 1976, Foundations of attribution: The perception of ongoing behaviour, in: J H Harvey, W I Ickes and R F Kidd, eds, *New directions in attribution research*, Lawrence Erlbaum, Hillsdale, NJ

Nisbett, R and Ross, L, 1980, *Human inference: Strategies and shortcomings of social judgement*, Prentice-Hall, Englewood Cliffs, NJ

Oodit, D and Simonis, U E, 1993, *Water and development: Water scarcity and water pollution and the resulting economic, social and technological interactions*; Wissenschaftszentrum Berlin für Sozialforschung, Berlin

Opotow, S and Clayton, S, 1994, Green justice: Conceptions of fairness and the natural world, Special Issue, Green justice: Conceptions of fairness and the natural world, *Journal of Social Issues* 50(3):1–11

Pawlik, K, 1991, The psychology of global environmental change: Some basic data and an agenda for cooperative international research, *International Journal of Psychology* 26:547–563

Pearce, F, 1994, Counting Africa's trees for the wood, *New Scientist* 142(1929):8

Platt, J, 1973, Social traps, *American Psychologist* 28:641–651

Rasinski, K A, Smith, T W and Zuckerbraun, S, 1994, Fairness motivations and tradeoffs underlying public support for government environmental spending in nine nations, *Journal of Social Issues* 50(3):179–197

Rayner, S, 1991, A cultural perspective on the structure and implementation of global environmental agreements, *Evaluation Review* 15:75–102

Rule, B G and Nesdale, A R, 1976, Moral judgement of aggressive behaviour, in: R C Green and E O O'Neal, eds, *Perspectives on aggression*, Academic Press, New York, NY

Schanne, M and Meier, W A, 1992, Risiko-Kommunikation: Ergebnisse aus kommunikationswissenschaftlichen Analysen journalistischer Umwelt- und Umwelt-Risiken-Berichterstattung, *Rundfunk und Fernsehen* 2:264–290

Scheler, M, 1926, *Die Wissensformen und die Gesellschaft Probleme einer Soziologie des Wissens Erkenntnis und Arbeit Eine Studie über Wert und Grenzen des pragmatischen Princips in der Erkenntnis der Welt Universität und Volkshochschule*, Neue-Geist-Verlag, Leipzig

Scheler, M, 1928, *Die Stellung des Menschen im Kosmos*, Reichl, Darmstadt

Schelling, T C, 1995, Intergenerational discounting, *Energy Policy* 23(4/5):395–403

Schlenker, B R, Britt, T W, Pennington, J, Murphy, R and Doherty, K, 1994, The triangle model of responsibility, *Psychological Review* 101:632–652

Seligman, M P, 1975, *Helplessness: On depression, development, and death*, Freeman, San Francisco, CA

Sias, P M and Jablin, F M, 1995, Differential superior-subordinate relations, perceptions of fairness, and coworker communication, *Human Communication Research* 22(1):5–38

Sigelman, C K and Waitzman, K A, 1991, The development of distributive justice orientations: Contextual influences on children's resource allocations, *Child Development* 62(6):1367–1378

Singer, E and Endreny, P M, 1993, *Reporting on risk – how the mass media portray accidents, diseases, disasters, and other hazards*, Russell Sage Foundation, New York, NY

Sjöberg, L, 1989, Global change and human action: psychological perspectives, *International Social Science Journal* 41:413–432

Sloan, T, 1992, Psychologists challenged to grapple with global issues, *Psychology International* 3:1,7

Sondak, H, Neale, M A and Pinkley, R, 1995, The negotiated allocation of benefits and burdens: The impact of outcome valence, contribution, and relationship, *Organizational Behaviour and Human Decision Processes* 64(3):249–260

Spada, H and Opwis, K, 1985a, Die Allmende-Klemme: Eine umweltpsychologische Konfliktsituation mit ökologischen und sozialen Komponenten, in: D Albert, ed, *Bericht über den 34 Kongress der Deutschen Gesellschaft für Psychologie in Wien 1984; Band II: Anwendungsbezogene Forschung*, Hogrefe, Göttingen

Spada, H and Opwis, K, 1985b, Ökologisches Handeln im Konflikt: Die Allmende Klemme, in: P Day, U Fuhrer and U Laucken, eds, *Umwelt und Handeln*, Attempto, Tübingen

Stern, P C, 1978a, The limits to growth and the limits of psychology, *American Psychologist* 33:701–703

Stern, P C, 1978b, When do people act to maintain common resources?, *International Journal of Psychology* 13:149–157

Stern, P C, 1992, Psychological dimensions of global environmental change, *Annual Review of Psychology* 43:269–302

Stokols, D, 1992, Environmental quality, human development, and health: An ecological view, Annual Conference of the American Psychological Association (1991, San Francisco, California), *Journal of Applied Developmental Psychology* 13:121–124

Tajfel, H, ed, 1982, *Social identity and intergroup relations*, Cambridge University Press, Cambridge, MA

Tajfel, H and Fraser, C, eds, 1978, *Introducing Social Psychology*, Penguin Books Ltd, Harmondsworth

Taylor, S E, 1989, *Positive illusions: creative self-deception and the healthy mind*, Basic Books, New York, NY

Taylor, S E and Brown, J D, 1988, Illusion and well-being: A social psychological perspective on mental health, *Psychological Bulletin* 103:193–210

Thomas, W I and Thomas, D S, 1928, *The Child in America*, Alfred A Knopf, New York, NY

Thompson, L and Loewenstein, G, 1992, Egocentric interpretations of fairness and interpersonal conflict, *Organizational Behaviour and Human Decision Processes* 51:176–197

Thompson, S C and Stoutemyer, K, 1991, Water use as a commons dilemma: The effects of education that focuses on long-term consequences and individual action, *Environment and Behaviour* 23:314–333

Tóth, F, 1994, Discounting in integrated assessments of climate change, in: N Nacicenovic, W D Nordhaus, R Richels and F Tóth, eds, *Integrative assessment of mitigation, impacts, and adaptation to climate change*, IIASA, Laxenburg, Austria

Tóth, F, 1995, Discounting in integrated assessments of climate change, *Energy Policy* 23(4/5):403–409

Tóth, F, 1997, *Die Rolle des Zeitfaktors im Management globaler Umweltveränderungen: Diskontierung und Kosten-Nutzen-Analyse*, paper presented at the WBL-Konferenz, October 29; Bonn

Tsai, Y M, 1993, Social conflict and social cooperation – simulating 'the tragedy of the commons', *Simulation and Gaming* 24:356–362

Turner, J C and Giles, H, eds, 1981, *Intergroup behaviour*, Blackwell, Oxford

Tversky, A and Fox, C R, 1995, Weighing risk and uncertainty, *Psychological Review* 102:269–283

Tversky, A and Kahneman, D, 1974, Judgement under uncertainty: Heuristics and biases, *Science* 185:1124–1131

Tversky, A and Kahneman, D, 1978, Causal schemata in judgement under uncertainty, in: M Fishbein, ed, *Progress in social psychology*, Erlbaum, Hillsdale, NJ

Tversky, A and Kahneman, D, 1983, Extensional versus intuitive reasoning: The conjunctional fallacy in probability judgement, *Psychological Review* 90:293–315

Tyler, T R, 1991, Using procedures to justify outcomes: testing the viability of a procedural justice strategy for managing conflict and allocating resources in work organizations, *Basic and Applied Social Psychology* 12:259–279

van Knippenberg, A and van Oers, H, 1984, Social identity and equity concerns in intergroup perceptions, *British Journal of Social Psychology* 23:351–362

Vermunt, R and Steensma, H, eds, 1991, *Social justice in human relations, Vol 1: Societal and psychological origins of justice; Vol 2: Societal and psychological consequences of justice and injustice*, Plenum Press, New York, NY

Vining, J, 1987, Environmental decisions: The interaction of emotions, information and decision context, *Journal of Environmental Psychology* 7:13–30

Wagstaff, G F, 1994, Equity, equality, and need: Three principles of justice or one? An analysis of 'equity as desert', *Current Psychology Developmental, Learning, Personality, Social* 13(2):138–152

Walster, E, 1976, New directions in equity research, *Advances in Experimental Social Psychology* 9:1–43

Wissenschaftlicher Beirat der Bundesregierung Globale Umweltveränderungen (WBGU), eds, 1993, *Welt im Wandel: Grundstruktur globaler Mensch-Umwelt-Beziehungen*, Economica, Bonn

WBGU, eds, 1994, *Welt im Wandel: die Gefährdung der Böden*, Economica, Bonn

Wicker, A W and King, J C, 1988, Life cycles of behaviour settings, in: J E McGrath, ed, *The social psychology of time*, Sage, Beverly Hills, CA

Wiener, J L, 1993, What makes people sacrifice their freedom for the good of their community, *Journal of Public Policy and Marketing* 12:244–251

Wit, A, 1994, *Provision of step-level public goods: effects of environmental and social uncertainty on group members' contributions*, paper presented at the EAESP small group meeting on social interaction and interdependence; Amsterdam, The Netherlands, April 28–May 1

Zimmerman, R, 1993, Social equity and environmental risk, *Risk Analysis* 13:649–666

8 Fairness and Local Environmental Concerns in Climate Policy

Shuzo Nishioka

Introduction – Why is Fairness to Local Concern so Important?

While too much emphasis is put on the global aspects of climate change, local concern and people's initiative as residents in local environment are not fairly dealt with, in spite of the fact that the actual impacts of climate change appear locally and those who respond to the impacts are the people who live there. Implementation for mitigating climate change consists of nothing but the accumulation of local activities. Current global responding strategy to climate change is based mostly on the mainstream valuing system, such as modern economics, which sometimes discards local autonomy, indigenous culture and tradition. Discussions so far (see IPCC Working Group III, 1996) dealt with equity matter mainly within the framework of relations between global and national (or country) aspect, not between global and local (free from 'national' context).

The environment is an integrated entity; only those who live in that place experience the synergistic benefit of nature there. Through their daily activities, local inhabitants foster their consciousness and respect of nature, and understand the advantage of living symbiotically with their surroundings. This wisdom has long been assimilated into their vernacular culture, tradition and institution. Their cognitive and evaluative abilities and wisdom, based on local experience of their environment, are collectively defined in this chapter as 'local ecological consciousness'. Ecological consciousness has high potential in effectively implementing climate policy, provided it is thoroughly integrated in the process. As illustrated here in the case of Asia–Pacific countries, the traditional story of the spirit living within nature is effectively working to maintain the sustainable use of natural resources, and native knowledge is used quite well to adapt to existing climatic disasters.

Why does ecological consciousness tend to be dropped out of consideration in the decision-making process of climate policy? Are there any failures in the decision process that cause unfairness in procedure and in climate change policy? How does this unfairness, if it exists, affect the efficiency of the climate policy? How can we harmonize local initiatives with efficient global climate policy?

This chapter, in its first half, illustrates the importance of local ecological consciousness in environmental policy by showing some examples in the Asia–Pacific region, and the effort of Asia–Pacific countries to involve their cultural values and traditional wisdom into a sustainable development path. In the latter half, this chapter discusses fairness considerations in global and local concerns in the case of global climate change policy and shows some examples of how local concerns become neglected in the decision process of climate policy. The last part considers how we can integrate local ecological consciousness fairly and efficiently into global climate policy.

ECOLOGICAL CONSCIOUSNESS IN ASIA–PACIFIC: ITS ROLE IN PRESERVING ENVIRONMENT AND ITS PERIL UNDER THE PRESSURE OF DEVELOPMENT

Local ecological consciousness is the basis of preserving local, regional and global environment. This consciousness is closely connected to people's attachment to nature, history and tradition in the very places where they live. But the contemporary overwhelming trend of globalization tends to destroy those local values and traditions through rapid industrialization, urbanization and unification of values on a global scale.

Asia–Pacific is the most rapidly developing region today, and traditional ecological consciousness is in peril since it is regarded as an obsolete value in the modern development pattern. But in the local community of the region, consciousness still survives and works to preserve local environment. The leaders of Asia–Pacific who are seeking new development patterns that are different from Western trends and unique to the region or the nation are now reexamining ecological consciousness as an alternative path.

Asia–Pacific: its unique dynamism in development and environment

Asia is now in the stage of rapid economic development with a large and growing population, accompanied by severe regional and global environmental problems.

A modelling study (AIM project team, 1997) predicts in its business as usual (BaU) scenario that in 2025 this region will have an urban population of two billion, 522 million passenger cars, a 50 per cent increase in demand for seafood from 1990, and a 36 per cent share of the global carbon dioxide emissions in the world, a substantial increase from the 25 per cent of today. The tropical deforestation rate is 1.2 per cent per year, soil degradation affects 10 to 50 per cent of the land area and 36 per cent of arable land is experiencing desertification. Asia has 72 per cent of the world's farming population but shares only 30 per cent of the world's arable land (Eco-Asia Project, 1997). This region's contribution to climate change is increasing, and it will be, at the same time, quite vulnerable to the impacts of possible climate change (IPCC, 1998).

These massive and dynamic developments of one region have never been observed in our history and are quite unique features of Asia–Pacific today. This region has a unique mixture of countries in different economic development stages – matured, developing and underdeveloped – and each country is connected to the other in culture and through economic relations. It became urgent to formulate policy that would integrate environment and development, and would combine regional or global policy with local concerns in the region.

Emergence of Asia–Pacific ecoconsciousness

In pursuit of sustainable development, traditional values and ways of life can provide the necessary philosophical underpinning. Unfortunately, many of these traditions have been unfairly dealt with, forgotten or disused in the course of recent economic development characterized by mass production and high consumption. There are indications, however, that interest in some of these traditions is reviving and that they may offer useful insights for formulating sustainable development policies.

To achieve sustainable development which integrates traditional and modern forces in society, it is necessary to establish a new pattern of development. The majority of people in the region are still poor and need economic development. On the other hand, environmental degradation has become apparent, even at this stage of development, so that integrating environmental protection and development is urgent.

In this process, it would be worthwhile for countries and people in the region to identify their own social and cultural traditions and practices, and examine the extent to which they can be incorporated within the new pattern of development. Joint efforts among the countries of the region to adopt this new approach towards development are now slowly beginning (Eco-Asia Project, 1997).

Local environmental consciousness in Asia–Pacific

There is a mix of wisdom, environmental ethos, education, institutional and religious beliefs, and lifestyles in tradition that enhances local environment and leads to a wise use of limited local resources. The sections below illustrate such local ecological consciousness found within Asia–Pacific countries (Eco-Asia Project, 1997).

Indigenous knowledge for the practice of sustainable agriculture

China is developing ecological farming throughout the country. This is a kind of farming practice which makes full use of energy and resources and uses little or no agrochemicals. Renewable energy sources such as biogas, wind, solar energy and hydropower are widely used. Through such farming practices, nutrients are fully utilized and recycled and virtually no waste is produced – thus the natural environment is protected while productivity is raised. Ecological farming is now practised in more than 100 places in China.

The traditional people of Indonesia's Siberut Island – the Mentanarians – have for centuries lived by traditional farming in the forested areas. Uniquely, they never use fire for land clearing since it is taboo. Instead, they immerse organic matter in swampy areas, so that a humidification process occurs through traditional technology. A widespread practice of using herbs, leaves and plants (from the neem tree) as pesticides instead of chemicals helps to control pollution in water bodies.

An example of a local self-sufficient production system in Sri Lanka

In Sinhalese and Tamil villages, all the commodities necessary for daily life, such as cereals, vegetables and spices, are produced within the living territory. A pond in the village is used for cultivating fish, orchards are planted around the village and clothes are made from the fibres harvested within the village. The production and material balance is restricted to the carrying capacity of the territory, and production activities minimize disruption of the environment. Recently, this living style is being affected by the export-oriented monocultural production system. However, in the 1980s, a policy was adopted to limit production of export commodities to individually specialized traditional goods that do not disrupt natural environmental systems.

Traditional education to maintaining precious soil in Tuvalu

As a small island state comprised of coral reef in the South Pacific, most of Tuvalu's soil consists of inorganic material and is inappropriate for agriculture. Small pits of organic soil studded in the island are the precious assets of Tuvalu farmers for planting taro and other major foods. In each houseyard in the country, people keep a 'taro pit' – a pit for planting their staple crops, filled with fertile organic soil inherited from their ancestors. Children are taught to put all the organic wastes, such as fallen leaves and household garbage, into this pit. In this way, a complete recycling scheme of nutrients, from soil to plants and plants to soil, is maintained. Government prohibits using chemical fertilizers in order to sustain productivity of the soil for the long term.

Cans of beer imported primarily from Australia do not fit this complete recycling system, and a huge pyramid of wasted beer cans can be found at the end of the long and narrow island, symbolizing the invasion of industrialization into this nature-harmonized country.

Traditional institutions and worship that advocate harmonious coexistence with nature

The Ambonese, an ethnic group in the Molluca Islands of Indonesia, possess a traditional custom of *sashi*. This is adopted strictly by the fishermen who are forbidden to go fishing in the sea at specific times of the year. They have observed this tradition from century to century. The scientific explanation of this *sashi* custom is that it provides ample time to maintain the reproductive and regenerative capacity of numerous species of fish. Without the *sashi* tradition, the fish stock could be depleted.

According to the religious and traditional customs of the Balinese, water is considered a sacred natural resource. Therefore, it is a sin to pollute the water. Moreover, the sustainability of water resources for agricultural activities is also ensured through traditional methods which result in efficient use.

Adapting to floods with traditional wisdom in Bangladesh

The low-lying areas of Bangladesh are flooded every year. Although sometimes the peak run-off of two big rivers, Brahmaputra and Ganges, coincides, causing serious damage as in 1988, flooding itself is not unusual in these areas. The flood carries fertile soil from upstream, and this helps cultivation of floating rice, soy bean and oil-producing plants. After the flood, rich fish ponds are formed. Bangladeshis live with and wisely use this periodical flooding. What is needed now is to control and ease the extremes of the flood.

These illustrations of existing practice suggest that the local consciousness, if fairly dealt with and well integrated into global-scale policy, can be a powerful concept and tool for responding to global environmental change.

Is ecological consciousness common to the region?

The Asia–Pacific region is quite diverse in every sense of the word. Nations and areas in the region are distinct from one another in terms of population size, customs, natural condition, level and speed of economic development, religion, culture, and tradition. At first sight, there seems to be no single pattern of behaviour, values or consciousness that is common to the whole region. Therefore, the illustrations above come from different backgrounds and cannot be understood within a common single context. It is difficult to find common characteristics regarding human relations with the environment that can act as guidelines for sustainable development. However, upon closer examination, some common features relating to values, modes of social decision-making and the relationship of people to nature can be discerned. These features are: Confucian ethic, Ramayana and Trip to the West (folklore stories in which monkeys play important role as heroes), similar language groups, rice culture, strong ties with the sea, and the importance of seafood, etc.

And we can find out some salient features to be passed on to future generations in terms of values and practices, modes of decision-making and relations between people and nature, such as: frugal lifestyles, sustainable agricultural and industrial practices based on indigenous knowledge and harmonious coexistence with nature.

Family ties are a strong unit of society. For example, in Thailand, many families live together traditionally from generation to generation in the same house or in the same village. The larger the family size, the less food and energy family members consume per capita due to the scale factor. Moreover, this lifestyle encourages respect for the locality and leads to environment-friendly landuse and natural resource use.

Major religions of the Asia–Pacific region emphasize frugality and simplicity, and teach the importance of meeting real needs rather than induced desire. In Pakistan, for instance, simplicity of dress and diet continues to be culturally and religiously practised. There are many vegetarians on the Indian continent. This way of life is environmentally friendly and tends to conserve natural resources.

Modes of social decision-making are another common feature in this region. The general tendency is consultation rather than confrontation, reliance on consensus in the decision-making process, reliance on settling disagreements through discussion rather than through legal action, pragmatic rather than doctrinal approach to problem solving, and informal rather than formal approach to governance.

At the empirical level, there are a number of trends evident in the development of many countries in the region that suggest a movement towards greater sustainability, reflecting the influence of Asian–Pacific ecoconsciousness. Economic growth is achieved at relatively low levels of energy use, and per capita caloric consumption of food is not high and is largely derived from plants rather than animals.

Local ecological consciousness as an alternative development path finder

The Asia–Pacific region has long been regarded as a latecomer in development. However, the geopolitical situation of the region changed drastically as Japan took the lead in economic development, followed by Asian newly industrialized economies (NIEs), such as Korea. With the vigorous development of the Association of South-East Asian Nations (ASEAN) countries, and nations such as China and India since the mid 1980s, the region has emerged as the growth centre of the global economy. At the same time, strong concerns have been voiced over the further expansion of the already huge population of the region, urban sprawl, and the continued contamination and degradation of the environment that could greatly hamper the future development of Asia and the Pacific, and negatively impact the world community, as in the case of climate change.

It is also recognized that many countries and regions have enjoyed less success in development but also face serious environmental problems, particularly with regard to deforestation, land degradation and the loss of biological diversity – all of which threatens the survival of the population's poorest. Therefore, the countries of the Asia–Pacific region have begun to search for a new path, shifting from conventional development patterns to sustainable development. Some of the countries in this region are already seeking this alternative path creatively through the fusion of modernity and tradition.

For example, the people of Singapore are largely descendants of immigrants from the Malay Peninsula, China, the Indian subcontinent and Sri Lanka. They have gradually acquired a distinct identity while retaining their traditional values, practices and customs. Some of the traditional values have influenced people's attitude towards the environment. Values such as thrift and respect are, for example, linked to care and proper use of resources. Wastage is thus frowned upon. To help strengthen national identity and to preserve the cultural heritage of different communities, the government has focused on five shared values. These are: nation before community and society above self; family as the basic unit of society; community support and respect for the individual; consensus rather than conflict; and racial and religious harmony.

CLIMATE POLICY: WILL IT WIDEN THE UNFAIRNESS BETWEEN GLOBAL AND LOCAL CONCERNS?

The global trends of contemporary society are working, rather, to increase unfairness towards local concerns. Globalization, through rapid development and expansion of information and transportation technology, and the formation of global markets that control material and money flows, established a decision mechanism based on a relatively narrow valuing system, eliminating local values which cannot work as common measures in the global market mechanism. Holistic values of small-scale, vernacular environments are abstracted into one-sided values that fit within the global mechanism, without consideration of local entities.

Does present global climate policy accelerate this trend of neglecting local concerns by dealing with them in an unfair manner? Does it weaken the potential power of local identity to protect local environments – which may lead to the alternative path in the long run? If it does, is this because of the intrinsic characteristics of climate policy?

Global change itself inherently tends to amplify existing vulnerability (IPCC, 1996), in particular in developing countries where life-support and living infrastructures are barely maintained through local efforts to adapt to unstable climate conditions. In the case of water resource distribution, for example, the predicted changing pattern of precipitation increases the number of people facing water shortage on the one hand, and at the same time increases the number of people with abundant water supply; consequently the existing inequity in water supply is widened throughout the world. The same situation happens in the case of sea level rise. Impacts are experienced by vulnerable sectors of society, such as small island nations and low-lying coastal cities; this can be the trigger which destroys the integrity of local societies and consequently the local environment.

Worldwide climate concerns and activated discussions, in general, enhance concern of the local environment. Although emphasis has been placed on the global concerns, there is increasing concern about the degradation of local environmental resources. Scientific efforts have been strengthened to focus on local environmental changes and a growing number of researchers are working with local surveys, under worldwide cooperative research programmes, such as the Land-Use and Land-Cover Change Research Programme and the Human Dimensions of Global Change Research Project. Echoing the growing interest in global environmental issues, and thanks to the development of visual media networking, the lifestyles of small towns and villages, biodiversity preserved in tropical forests, and local practices of agricultural production have easily been introduced to the world, and many people now appreciate the disparate values of differing cultures.

However, some aspects of the global climate policy, which unfairly neglects issues of locality and local concerns, work against preserving local environmental consciousness. Explanations for this come from essential characteristics of the climate issue and from the same driving forces which govern the present globalized world. The following sections include some aspects of this mechanism.

Mitigation versus adaptation – unfairness in participation

In the series of decisions made by the Conference of the Parties, the Global Environmental Facility (GEF) fund was established mainly for mitigation measures, but also in a limited area for adaptation. While the identification and assessment of options for mitigation are included in activities eligible for assistance from GEF funding, the options for adaptation are only partly funded for stage I activities, such as planning, which includes studies of possible impacts of climate change to identify particularly vulnerable countries or regions and policy options for adaptation and appropriate capacity-building. Stage II measures, including further capacity-building, which may be taken to prepare for adaptations, and stage III measures to facilitate adequate adaptation, including insurance and other adaptation measures, are not approved as eligible.

Every participating country should have the capacity to respond to climate change in its own way. For countries who produce few emissions but are supposed to suffer severe impacts, the only way forward is to have firm measures which they can manage autonomously by themselves, not depending on measures, such as emission control, which are in in other stakeholders' hands.

It is, no doubt, reasonable to give priority to mitigation; if we prevent climate change, preparation for adaptation will be unnecessary. Nevertheless, local responding initiatives should be dealt with equally, such as emission control, from the procedural fairness point of view.

Migration versus residence

In the speech of the Third Conference of the Parties to the United Nations Framework Convention on Climate Change (UNFCCC), the prime minister of Tuvalu, a small island state in the South Pacific, threatened by sea level rise, strongly emphasized how his island is precious and unique to the people who live there.

> *While parties of UNFCCC here in Kyoto debate over what emission reduction to take, Tuvalu continues to bear and suffer the increasing cost of climate change impacts, which is threatening the very existence, culture and unique identity of Tuvalu as a member of the global community. There is nowhere else on Earth that can substitute for our God-given homeland in Tuvalu. The option of relocation as mooted by some countries is, therefore, utterly insensitive and irresponsible.*

From the global economic point of view, the cost of migrating less than 10,000 people of Tuvalu to other countries will be far less than the cost of reducing a percentage of the emissions in a big industrialized country. But in this case, to the residents of Tuvalu, the value of their island far outweighs the figure it has been given in the contemporary economic evaluation system. The local ecological value as a whole is not fairly valued in this case.

Native knowledge versus modern technology

Climate matters are to be solved by the accumulated effort of local activities. Local people know well what technology is most applicable and efficient. They have a profound knowledge of neighbourhood ecosystems and use local environments wisely. However, sometimes indigenous knowledge is treated lightly and neglected by the industrialized world.

A New York based company extracted medicine from the neem tree of India and applied for a patent. But Indian farmers have used the seeds of this tree as a pesticide for centuries, and so the application is already obvious and unpatentable (*Science*, vol 269, 15 September 1995). The intellectual property system established for the modern industrialized world does not always take local knowledge into consideration.

A similar situation happens when climate policy gives priority to adaptation by modern technology over local knowledge. In order to adapt to sea-level rise, rubble work that preserves coral reefs should be used rather than concrete breakwaters. Adaptation by local wisdom should be respected most and capacity-building for autonomous adaptation by local people should be encouraged. In this case, there is a biased recognition of the capability of local initiatives.

Dropping out in the process of integrating information – deficiency of decision-making tools

It is difficult to create an effective quantitative communication tool to understand local situations and concerns. For example, the integrated assessment model (IAM) is widely used as a powerful decision-making tool for formulating climate policy. The modellers have to deal with wide regions and fields related to climate science, climate change impacts and responding strategy. It is out of their capability to incorporate detailed local concerns within their model. Very few IAMs shown in IPCC reports are the bottom-up type that emphasize local responding technologies, and there remains large uncertainty in evaluating local impacts and the effect of adaptation measures. Lump-sum monetary terms are used to describe cost-benefit analysis in the model, which reflects little local ecological value.

So far, no scientific tools and processes are established to reflect local concerns of global policy decisions, and sporadic information by local non-governmental organizations (NGOs) is the only way to reflect local value and concerns. In this case, the deficiency of premature information systems to serve as decision-making processes is highlighted.

Marketing value versus non-marketing value

An example of this case was in the heated argument in the IPCC's *Second Assessment Report* on the application of cost-benefit analysis. The value of a human life can be estimated as the amount paid to his bereaved family upon accidental death – so modern economics teaches. Based on such calculations, the value of human life in developing countries is estimated far less than in developed countries, because of the different economic situation. The Tuvalu case mentioned above can also be included in this category. Attachment to the family or homeland can be quantified by modern economics in a limited degree only and can cause a sense of inequity if applied naively.

Monovalue versus multivalue – neglecting holistic existence

The global climate policy has created a new common currency – carbon equivalent or GHG equivalent – by which local environments can be evaluated. Existing nature is valued simply in terms of tonnes of carbon dioxide, traded in the global virtual market, and forests are managed only as reservoirs of carbon. In this process, all other local functions and values of natural forest are discarded.

Tropical forests have many other values, such as market value as a timber source, option value as undiscovered genes, non-market value as the capacity to control watersheds, and the preservation of biodiversity. The decision in the Kyoto Protocol to use a 'sink' for reduction measures to attain the national target, and to establish global common market mechanisms, such as joint implementation and clean development mechanisms, tends to make forest managers increase the volume of trees, neglecting other functions which are attributed to forest ecosystems.

In this case, global climate policy does not value fairly local, vernacular environments – and by stressing open markets in trading carbon, climate policy destroys holistic values of the local environment.

Unfairness in accessibility to scientific information

Climate change issues are based on knowledge in diverse scientific fields. Accessibility to scientific information and understanding of the characteristics of climate change phenomena are so vital to correct decision-making in highly strategic international negotiations. But very little site-specific information is provided to the international decision-making process and little information, in turn, is transferred to the local level. One reason comes from lack of scientific studies on the local level, restricted by limited scientific knowledge, such as in predicting local impacts by high-resolution climate prediction models. Another reason is the lack of field study in local environments, especially in developing countries, because of the shortage of research capacity.

In the IPCC's *Special Report of Regional Impacts* (1998), research results on developing countries are quite limited in comparison to results on developed countries. Participants from developing countries are fewer in number, and the knowledge of climate change is not well transferred from local level to global level and vice versa.

Without enough material for decisions that may affect local environments, policy-makers on the local level are forced to participate in global-scale agreements in climate policy. Hence, accessing information on complex scientific-political relations differs unfairly between developed and developing countries.

Conclusion: Fair and Efficient Climate Policy Towards Long-Term Sustainability

Some of the local ecological consciousness fostered within indigenous traditions plays a big role if well integrated within climate policy. This ecoconsciousness, which ultimately means establishing a harmonious existence between local society and nature, is not only inherent in the Asia–Pacific region, but is common throughout the world, even though its features diverge from place by place. Ecoconsciousness, however, is being diminished by the intrusion of contemporary industrialized development patterns within local community, though some Western traditions have been reformed through, for example, citizens' movements or folksongs and tales.

Local ecological consciousness can be the basis of an alternative path, for developing countries, towards a sustainable future. The premise of a linear development pattern, that every developing country aspires to the same goal by following the same track of their predecessor, is to be reconsidered within the global context of sustainability. Some Asia–Pacific countries are establishing their own development plan based on their traditional values, as is seen in the case of Singapore. Integrating these local values and movements within worldwide climate policy by enhancing local capacity is what must be discussed when establishing global collaboration mechanisms.

The overwhelming pressure of globalization is diminishing holistic values of local environments and local initiatives by setting simple tradeable values in the global market. Present global climate policy accelerates this tendency by unfairly considering the local community in procedures and consequences. The bias towards mitigation deprives local communities of their own responding measures, which leaves vulnerable regions without any autonomy and their destinies in the hands of mitigation groups. Unequal opportunities to access the necessary scientific information and insufficient information on local level situations tend to give low weight to local initiatives.

Considering the potential role that local ecological consciousness has to play, the following concerns should be positively integrated within the global climate policy.

- Each country should rediscover and identify traditional elements which conserve the environment, should protect and foster these elements, and all countries should respect each other's unique characteristics of ecological consciousness.
- Since local ecological consciousness is so weak against the overwhelming pressure of present globalization, some measures must be taken to protect it, from the Rawls equality point of view.
- International cooperation mechanisms, such as tradeable permits, joint implementation and clean development mechanisms, should be carefully designed to protect the holistic values of local environments, avoiding single value criteria.
- Local responding capacity, especially adaptation to climate change, should be enhanced by fostering local consciousness and knowledge, as well as by transferring technology that considers local adaptability.

- Information gaps between global and local levels should be narrowed by scientific research and educational programmes on local environments and by strengthening the communication path between global policy decision and local people.
- Climate policy, which is but a part of international collaborating processes towards sustainable development, should be integrated with local and national strategies that respect local initiatives to protect the environment in situ.

REFERENCES

AIM Project Team, 1997, *Asia–Pacific Integrated Assessment Model*, National Institute for Environmental Studies

Eco–Asia Project, 1997, *A Long-Term Perspective on Environment and Development in the Asia–Pacific Region*, Environment Agency of Japan

IPCC (Intergovernmental Panel on Climate Change), 1996, *Climate Change 1995: Economic and Social Dimensions of Climate Change*, Contribution of Working Group III to the Second Assessment Report of the IPCC, Cambridge University Press, Cambridge

IPCC (Intergovernmental Panel on Climate Change), 1998, *The Regional Impacts of Climate Change: An Assessment of Vulnerability*, Cambridge University Press, Cambridge

Science, 1995, Patent on native technology challenged; **269**:1506, 15 September 1995

9 Justice, Equity and Efficiency in Climate Change: A Developing Country Perspective

P R Shukla

Justice and Equity in Climate Change

'Justice', as Rawls (1971) has pointed out, 'is the first virtue of social institutions'. Justice principles are needed to evaluate or propose alternative distributions. Justice in this sense is a distributive concept. A distribution may affect the evaluation criteria, such as welfare, directly or indirectly. Equity refers to normative criteria for judging the distribution. The equity is also defined as 'the quality of being fair and impartial' (see Flexner, 1987). In either sense, equity is basic to the justice process. The global climate-change phenomenon, arising from the accumulation of greenhouse gases in the atmosphere emitted by anthropogenic activities, influences the welfare globally. The climate-change mitigation regime requires evaluation of alternative policies that would redistribute the welfare effects among nations. It is therefore an important issue requiring justice intervention.

The complexity of the justice question, in the context of climate change, arises from the global and long-term character of the problem and the asymmetry of actions and their external effects (impacts) spatially and temporally. The actions causing the climate change – the greenhouse gas emissions – take place globally. Limiting emissions helps to mitigate the impacts. The emissions limitation is a justice problem requiring distribution of rights to emit – to use the atmosphere – to different nations. This is just one dimension of the justice problem. Adaptation to impacts and compensation to impacted parties are the other justice concerns.

Climate change impacts have two characteristics: for a given global emissions trajectory, the distribution of impacts across the nations is independent of the emissions profile of each nation; and the impacts are felt over a long time horizon due to the long life of greenhouse gases in the atmosphere (see Houghton et al, 1996). The justice has to therefore address both the intragenerational and intergenerational equity concerns. This chapter discusses primarily the distribution of emissions entitlements across nations, an intragenerational distributive justice problem, and the efficiency of the distribution, from the perspective of developing countries.

UNFAIR BACKGROUND CONDITIONS

What shapes the distinct justice concerns of developing countries in the climate change issue? This is a practical question, the understanding of which may help to lay a robust foundation for a climate mitigation regime. The primary justice issue in the present climate negotiations pertains to the distribution of emissions entitlements. This is a bargaining problem with multiple players. In bargaining theory, players reach a voluntary agreement only when it makes every player better off (the Pareto improvement) compared to the status quo (see Kverndokk, 1995). When bargaining power is unequally distributed, the agreement may not be Pareto optimal.

The current balance of power is unequally distributed in favour of developed countries – who controls most of global capital, military power, natural resources and knowledge resources. The unfair background conditions are the end-products of historical and natural processes. Their influence on the rules of bargaining is the principle justice concern for developing countries.

Developed nations are the main contributors to greenhouse gas emissions historically. Their emissions are now declining, whereas emissions from developing countries are rising. The energy resource endowments in many developing countries are more polluting, such as coal in China and India. The unfavourable background conditions, compounded by rising future emissions, can potentially translate into an agreement that may transfer mitigation burdens to developing countries in contravention with several equity criteria discussed later. Justice in this context is vital for inviting wide participation from developing countries in the climate regime, a main criteria for the success of the regime (see Kverndokk, 1995).

ASYMMETRY OF EMISSIONS AND IMPACTS

The causal relationship of emissions with impacts is central to the climate change issue. The climate change burden includes the costs of emissions mitigation, adaptations, impacts and risks (see Chichilnisky and Heal, 1993). The asymmetry between emissions and impacts highlights the equity concerns in the climate change problem, since a greater burden of impacts is distributed to poorer nations by natural processes, while most anthropogenic greenhouse gas emissions arise from economic activities in affluent nations. Since the impacts are inadequately understood, the higher risks are imposed on poorer nations. The valuation of impacts also poses serious difficulties, especially in developing countries where the insurance markets are undeveloped and the valuation is plagued by the controversies of value of life and future purchasing-power parity trajectories (see Shukla, 1996b).

The climate change phenomenon coincides with the period when many developing countries are set for rapid economic growth. The timing of the phenomenon is thus unfavourable to developing countries. The asymmetry of emissions and impacts, and the unfavourable timing, together make the climate change a classic problem requiring balance among economic development, environment and distributive justice. The climate-change mitigation strategies thus transgress into a larger economic development agenda, especially in developing countries, requiring simultaneous examination of justice, equity and efficiency concerns.

MINIMIZING THE BURDEN SIZE

The United Nations Framework Convention on Climate Change (UNFCCC) advises that *policies and measures to deal with climate change should be cost effective so as to ensure global benefits at the lowest possible cost* (see UNFCCC, 1992, Article 3.3). The UNFCCC also exhorts that: *parties that would have to bear a disproportionate or abnormal burden under the convention should be given full consideration* (see UNFCCC, 1992, Article 3.2). The aim of climate negotiations is thus to minimize the cost as well as to distribute the burden equitably. In other words, the UNFCCC is concerned with minimizing welfare burden, and minimizing the mitigation cost as a corollary.

The economic theory of this perspective, the neoclassical economics, assumes existence of efficient market dynamics universally. The market efficiency and cost effectiveness are hence equivalent. To be efficient, the mitigation actions across nations, sectors and time periods must have equal discounted marginal costs. Since climate change is a global and long-term problem, the search for efficiency leads naturally into a 'where and when' flexibility (see Richels et al, 1996) – that is, to decide the location and time of mitigation actions which equalize the marginal costs across the nations and in time, and thereby minimize the global mitigation cost (the size of the burden). Neoclassical economics, and the agenda of the developed countries, goes only this far. Distributing the mitigation burden is considered a separate problem, merely a secondary side-payment issue.

However, the emissions mitigation cost is just one component of the climate change burden. The others are costs of impacts, which are distributed across nations by climatic processes, and costs of adaptation. The aim of climate negotiations is to minimize the welfare losses and not the emissions or mitigation costs alone. Minimizing the welfare losses requires dealing up front with equity – that is, the distribution of total welfare burden, including the distribution of side payments. This is the basic aspect of developing country perspective in climate negotiations. The welfare concept subsumes both the efficiency and equity. Climate negotiations should first address the right measure for the burden. Economics is only then needed to help minimize the size of this burden.

SEPARATING EQUITY AND EFFICIENCY

A principal objective of the global climate change regime is to decide the norms for using the atmosphere, a global common. The Coase theorem (see Coase, 1960) stipulated that, in absence of transaction costs, the market exchange will lead to an efficient resource allocation regardless of the distribution of rights.

Alternatively stated, the Coase theorem suggests that the process of minimizing the size of the burden is independent of the burden-sharing scheme. This result means that market efficiency and equity are separate issues. To many neoclassical economists, this is a sufficient justification to discard equity concerns from the domain of economics altogether.

This perspective, which during the past decade has gained ground with the emergence of the new world economic order, has treated the climate change problem merely as a search for a globally efficient mitigation regime. The focus of the mitigation debate is restricted to minimizing the size of the burden. Market tautologies, such as the equalization of marginal costs across nations, sectors and time periods, have thus emerged as the sole means of deciding the participation of each nation in mitigation. The choice of efficient market instruments, such as emissions permits or taxes, has been made the principle agenda for global negotiations. This perspective, which justifies ignoring equity altogether, suits well the interests of industrialized nations in the climate change problem.

EQUITY AND EFFICIENCY: A DEVELOPING COUNTRY PERSPECTIVE

The above perspective has apparent limitations. In negotiations, parties not only have cooperative needs to minimize the global burden, but also competing needs to minimize own share. The game theoretic approaches which are designed around the concepts of efficiency and bargaining positions have contributed little to the global burden-sharing problem, which is primarily a justice issue. The market efficient-mitigation solutions naturally lead to significant mitigation actions in developing countries (see Manne and Richels, 1996; Richels et al, 1996) since the opportunities for low cost mitigation there are far greater due to prevailing market inefficiencies and inadequacies. Such proposals, arrived at by applying the efficiency concept alone, have raised obvious equity concerns from developing countries in the absence of simultaneous burden-sharing proposals (see Shukla, 1996a).

Developing country perspectives make a contrary interpretation of the Coase theorem compared to the existing neoclassical view. The cost effectiveness, à la Coase theorem, is a simple technical issue of finding and agreeing upon a market instrument such as emissions permits. The substantial issue in global negotiations, then, is not efficiency but distribution or equity. Since the stakes are high, the interests of bargaining parties are in conflict and their justice perceptions differ widely; the principal challenge before global negotiators is not to find an efficient instrument, but to harmonize these diverse perceptions and to arrive at a widely acceptable mitigation arrangement.

Justice, from a developing country perspective, is thus the primary concern of climate-change mitigation negotiations. Its aim is to arrive at a 'fair and impartial' distribution of the mitigation burden. The developing country position on the neo-classical approach to the climate change problem can be succinctly summed up by reiterating the observation made by Rawls (1971):

A theory, however elegant and economical, must be rejected or revised if it is untrue; likewise laws and institutions no matter how efficient and well arranged must be reformed or abolished if they are unjust.

An efficient world order for climate change mitigation, if unjust, would need alterations.

Equity in the UNFCCC

Varied equity criteria are explicitly considered in the principles of the UNFCCC (see UNFCCC, 1992). The special consideration for developing countries, as articulated through the 'common but differentiated responsibilities' (Article 3.1) clause, requires leadership (greater acceptance of burden) by developed countries in combating climate change. Strong equity concerns are also reflected in the special attention and considerations proposed for developing country parties who are particularly 'vulnerable to the adverse effects of climate change' and for those who have 'to bear disproportionate or abnormal burden under the convention' (Article 3.2). The exclusion of developing countries from any binding emissions-limitation commitments in the present negotiations culminating in the Kyoto Protocol (see UNFCCC, 1997) reaffirms these equity concerns enunciated in the UNFCCC.

Procedural and consequential equity

Equity concerns are of two kinds (see Banuri et al, 1996): the procedural and the consequential. Procedural equity refers to 'impartiality and fairness' in the process of delivering and administering justice. Consequential equity relates to assessing and remedying the consequences arising from climate change and the mitigation actions. Principles such as the participation of the affected parties in the justice procedure, or equal treatment of all before the law, belong to the notion of procedural equity. The specific procedural equity problems for developing countries in the global climate-change regime arise from their poor information base, weak bargaining strength and inferior capacity to deal with climate change.

Consequential equity addresses the sharing of the climate change burden. Various approaches to consequential equity such as parity, proportionality, priority, utilitarianism and distributive justice exist (see Banuri et al, 1996). However, there is no consensus on the superiority of any single approach. The concerns relating to consequential equity from the developing country perspective arise from:

- their low historical contribution to the existing stock of greenhouse gases in the atmosphere (see Table 9.1);
- their very low per capita emissions which are only a fraction of that in developed countries (see Banuri et al, 1996);
- high risk from climate change impacts (such as small island nations) compared to the size of their economy; and,

Table 9.1 Historic CO_2 and methane contribution by region, 1800–1988 (in percentages)

Region	Industrial CO_2	Total CO_2	$CO_2 + CH_4$
1 OECD North America	33.2	29.7	29.2
2 OECD Europe	26.1	16.6	16.4
3 Eastern Europe	5.5	4.8	4.7
4 Former USSR	14.1	12.5	12.4
5 Japan	3.7	2.3	2.3
6 Oceania	1.1	1.9	1.9
7 China	5.5	6.0	6.3
8 India	1.6	4.5	4.8
9 Other Asia	1.5	5.0	5.2
10 North Africa and Mid-East	2.2	1.7	1.8
11 Other Africa	1.6	5.2	5.2
12 Brazil	0.7	3.3	3.3
13 Other Latin America	3.2	6.5	6.5
Developed countries (1–6)	83.8	67.8	66.9
Developing countries (7–13)	16.2	32.2	33.1
World	100.0	100.0	100.0

Source: Grübler and Nakicenovic, 1991

- lack of resources, technologies and capabilities to mitigate the impacts.

The rest of the discussion in this chapter focuses on these consequential equity concerns.

EMISSIONS FROM DEVELOPING COUNTRIES

The current per capita emissions (see Figure 9.1) and historical contributions of developing countries to the existing emissions stock in the atmosphere are small compared to industrialized nations (Table 9.1). It is argued, however, that the rapidly rising emissions from developing countries will reverse this position in a few decades. While the emissions from developing countries are growing at rates higher than those for developed countries, the per capita emissions of developing countries will continue to remain substantially lower throughout the next century in the absence of climate change interventions. This is evident from the analysis of per capita emissions for three well-known emission scenarios proposed by the Intergovernmental Panel on Climate Change (IPCC), namely the IS92a, c and e scenarios (see Table 9.2). These scenarios assume no climate change interventions and cover a full range of possible future emissions (see Leggett et al, 1992; Pepper et al, 1992; Alcamo et al, 1995).

For each scenario, the per capita emissions in developing countries will be below a third of the developed country emissions even in the year 2100 (Table 9.2). It is likely that some developing countries may reach the emissions level of developed countries before the end of the next century. Yet, even after a

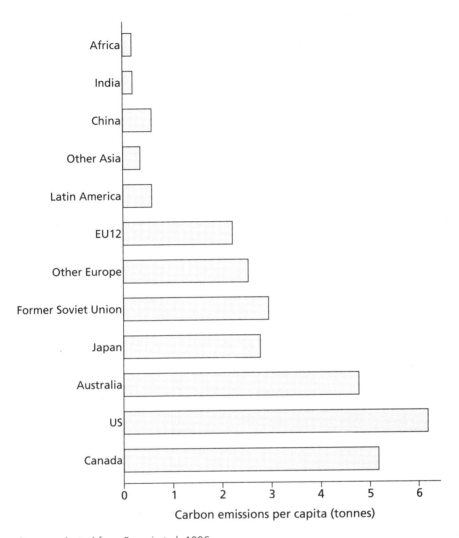

Source: adapted from Banuri et al, 1996

Figure 9.1 Carbon emissions per capita – 1993

century, the per capita emissions in most developing countries will remain far below those in industrialized countries. While the income gap is expected to narrow, the per capita incomes in developing countries will remain a fraction of developed country incomes throughout the next century. Hence, the population of most developing countries will bear only a marginal mitigation burden during the next century under a welfare-loss minimization regime.

Future global emissions and mitigation needs

The IPCC emissions scenarios suggest that in the absence of climate change mitigation interventions, the emissions during the next century will result in

Table 9.2 Per capita emissions (tonne of carbon per year) for IS92a, c and e

Region	Year	IS92a	IS92c	IS92e
Developed countries	1990	3.63	3.63	3.63
	2025	4.20	3.28	5.11
	2100	4.87	1.96	7.93
Developing countries	1990	0.39	0.39	0.39
	2025	0.73	0.53	0.95
	2100	1.36	0.57	2.49

very high concentrations of greenhouse gases by the end of the century (see Wigley et al, 1997). The IS92a scenario, often considered a mid-emissions scenario, would cause the carbon concentration from the fossil emissions alone to exceed 700 parts per million volume (ppmv). The high emissions IS92e scenario would lead to an alarmingly high concentration of 950 ppmv. In these scenarios, the rising emissions from developing countries are a substantial fraction of global emissions. Thus, the concentrations will reach alarming levels even if developed countries drastically reduce their emissions throughout the next century. Developing country participation, therefore, will be essential to keep the 'greenhouse gas concentrations in the atmosphere at a level that would prevent dangerous anthropogenic interference with climate system' – the ultimate objective of the UNFCCC. The global climate policy needs to decide upon acceptable stabilization levels to minimize the total climate burden, as well as upon an equitable burden-sharing arrangement.

The cumulative emissions for the period 1991 to 2100 for different IS92 emissions scenarios are shown in Table 9.3. Two well-known stabilization trajectories are those proposed by the IPCC (see Schmiel et al, 1995), called 'S' trajectories, and by Wigley et al (1996) called 'WRE' trajectories. The cumulative emissions for these trajectories are shown in Table 9.4. A comparison of emissions from these trajectories with IS92 emissions suggests that stabilization at 750 ppmv will require substantial mitigation throughout the next century. Cost effectiveness apart, the substantial mitigation needs will necessitate the early participation of developing countries. The pertinent question in the negotiations, therefore, is not whether or when developing countries will join the regime, but how the mitigation burden will be shared. A just burden-sharing regime is vital to ensure wide participation from developing countries, which is a necessary condition for cost effectiveness and stabilization of concentrations.

EQUITY PERSPECTIVES AND PRINCIPLES

Diverse equity perspectives and principles are debated in the context of climate change (see Rose, 1990; Rose and Stevens, 1993; Banuri et al, 1996). The central ones are:

- per capita entitlements (see Grubb, 1989; Agarwal and Narain, 1991), or egalitarian principles (see Rose, 1990);

Table 9.3 Cumulative carbon emissions for the period 1991–2100 for IS92 emission scenarios

Emission scenario	Cumulative emissions: 1991–2100 (giga tonne carbon)
IS92a	1500
IS92b	1430
IS92c	785
IS92d	975
IS92e	2187
IS92f	1845

- historical responsibilities (see Hyder, 1992);
- basic needs;
- obligation to pay – a composite criterion that combines historical responsibility and basic needs (see Hayes, 1993);
- Rawlsian criteria (see Benestad, 1994);
- ability to pay (see Smith et al, 1993);
- 'grandfathered' emissions (see Bodansky, 1993); and
- utility maximization (see Chichilinsky and Heal, 1994).

The present emissions limitations negotiations, leading to the Kyoto Protocol, have centred on the 'grandfathered' emissions criterion. The mitigation burden is distributed among developed nations as a percentage reduction from their current emissions. Developing countries are excluded from the binding commitments, keeping in mind the historical responsibility and ability to pay criteria.

The 'grandfathered' emissions criterion gives higher entitlements to present polluters. This distribution inherently disfavours developing countries, since their current emissions are very low but are rising rapidly. Since developing countries are excluded for the present from binding commitments under the Kyoto Protocol, the grandfathered criterion is not contested. The need in future to move away from the grandfathered criterion towards equal per capita emissions is articulated through the 'convergence' framework. This framework proposes to begin, at present, with grandfathered emissions entitle-

Table 9.4 Cumulative carbon emissions (1991–2100) for 'S' and 'WRE' emissions trajectories for stabilization of CO_2 concentrations

| Stabilization target | Cumulative carbon emissions[a] (giga tonne carbon) | |
	'S' trajectory	'WRE' trajectory
450 ppmv[b]	628	714
550 ppmv	872	1043
650 ppmv	1038	1239
750 ppmv	1194	1348

a Cumulative emissions vary due to different emissions trajectories (Wigley et al, 1996)
b ppmv: parts per million volume

ments, which in future can converge to equal per capita entitlements that can match the desired stabilization trajectory. While the convergence framework is a practical negotiation instrument, equity considerations are still needed to determine the 'convergence target' level and the timing of the entry of developing countries into the protocol.

Equity and the convergence criteria

The convergence framework proposes to bridge the gap in per capita emissions between developed and developing nations within a few decades. *Prima facie*, the compromise appears favourable to developing countries and also balanced for developed countries. The grandfathered emissions criterion suits well the present needs of developed nations. Equally, per capita emissions entitlements fit well with developing country demands. The convergence framework has limitations and unless it is designed with strong equity concerns, it will contravene several accepted equity criteria such as the obligation to pay, Rawlsian criteria and utility maximization.

The per capita emissions gap between developed and developing nations today is manifold (Figure 9.1). Under business as usual, the trajectories of many developing countries will cross the target, thus making them the net buyers of entitlements after a few decades. An early entry of developing countries into a convergence protocol which uses grandfathered allocations in the early decades and equal per capita emissions later on can be doubly inequitable. This is likely to happen under the conventional conception of convergence (see Jepma and Munasinghe, 1998), illustrated in Figure 9.2a, unless the entry of developing countries and their entitlement allocations are decided based upon equity criteria such as historical responsibility, ability to pay and Rawlsian philosophy. The presumption that the per capita emissions trajectories of developed and developing nations will converge without crossing the target makes developed nations the net gainers of emissions entitlements (Area A: Figure 9.2a) and developing countries the net contributors (Area B: Figure 9.2a) for past and future.

The income effect (see Kuznet, 1955) is likely to cause the per capita emissions from developed countries to shift to a declining trajectory earlier than in developing countries. In later periods, developing countries will experience a rising burden of buying the entitlements or incurring mitigation costs. For instance, the present per capita carbon emissions from China are 0.6 tonnes of carbon. These are rising rapidly. The stabilization trajectories (see Wigley et al, 1997) require the convergence to be below 0.7 tonnes of carbon per capita later in the next century. Even if the targets in earlier periods are set at higher levels, say at around one tonne of carbon per capita, China will become a net buyer of entitlements within a decade or will have to incur significant emissions mitigation costs for shifting to a lower carbon emissions trajectory. This can be unfair from the historical responsibility and ability-to-pay perspectives.

A more equitable convergence scheme, from a developing country's perspective, may follow the trajectories as shown in Figure 9.2b. The developing country per capita emissions can, firstly, cross and then converge to the target level, while on a downward path following the Kuznet curve.

a Conventional convergence concept

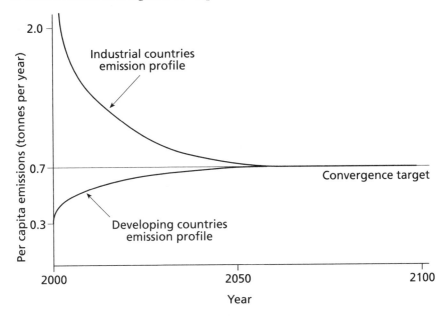

b Just and equitable convergence profiles

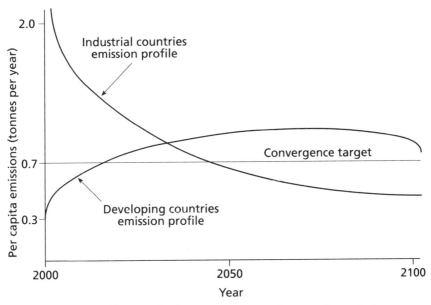

Figure 9.2 Trajectories for convergence of per capita emissions

Alternatively, the equity criteria would recommend a changing target level that at first follows a rising and then a declining trajectory, together with an early entry of developing countries so that they earn and bank the entitlements for later use.

Technology protocol with income-based entry

Edmonds and Wise (1997) propose a different equity criterion that considers an income-based entry for developing countries into a technology-oriented mitigation protocol. The protocol is proposed in three parts:

(1) Any new fossil fuel electric power capacity in Annex I nations installed after the year 2020 must scrub and dispose of the carbon from its exhaust stream.
(2) Any new synthetic fuels capacity must capture and dispose of carbon released in the conversion process.
(3) Non-Annex I nations which participate must undertake the same obligations that Annex I nations undertake when their per capita income, measured by purchasing power parity, equals the average for Annex I nations in 2020.

The global analysis based on the model shows that the protocol is adequate to stabilize the CO_2 concentrations below the 550 ppmv level. Developed countries (Annex I) join the protocol after 2020, but their effective contribution in the analysis begins after 2035 due to, firstly, the slow change in technology stock due to lower growth rates and, secondly, long time periods used for the model analysis. The additional cost of carbon removal technologies is thus small in early decades. As per part 3 of the proposal, most developing countries would enter the protocol during the second half of the next century. *Prima facie*, the income clause in the protocol seems fair to the developing countries. In reality, however, developing countries such as China enter the protocol around 2040 – just a decade after developed countries have begun implementing the protocol.

While the protocol demonstrates the feasibility of stabilizing the concentrations through a strong technology push, it is neither cost effective nor does it ensure an equitable distribution of the mitigation burden. The merit of the protocol is that it directly integrates technology strategy with the equity criteria and the mitigation strategy. The protocol has strong technology assumptions which may not be attainable at reasonable costs. The sense of security promised by the results that a strict stabilization trajectory, such as 550 ppmv, can be attained by mitigation programmes which can begin as late as 2030 to 2040 may prove false if these strong technology related assumptions do not materialize.

The income-based 'graduation clause', which decides the entry of developing countries within the protocol, is an indirect equity measure. Since the rest of the technology conditions are presumed identical for all entrants, the protocol gives a sense of 'fairness', which is illusive since nations have different economic structures, resource endowments, technological capabilities and emissions profiles.

CONCLUSIONS

The three important attributes, the long time horizon, global nexus and the need for wide participation, make climate change mitigation a truly complex problem. Success of the regime depends on the robust foundations built early on (see Toman and Burtraw, 1991; Kverndokk, 1995; Rayner and Malone, 1998). Justice, equity and efficiency are the three pillars of the climate mitigation regime. While the efficiency concerns are amply addressed in the climate debate, justice and equity have received meagre attention. The future emission scenarios developed by the IPCC suggest that the ultimate objective of the UNFCCC – the stabilization of greenhouse gas concentrations – cannot be achieved without participation of developing countries in mitigation. The economic efficiency studies also indicate that the global cost effectiveness will require sizeable emissions mitigation and adaptation actions in developing countries. Participation of developing countries in the climate regime is therefore necessary and cost effective.

The basic concern of developing countries is not whether or when to initiate the mitigation actions, but how the mitigation burden will be distributed among nations. This is a justice issue, concerned with an equitable distribution. The unfavourable bargaining position has contributed to myriad apprehensions about the procedural and consequential equity in negotiations. Concerns on the matter arose in developing countries even before the Framework Convention was drafted (see Agarwal and Narain, 1991). Climate-mitigation policy studies have shown that the economic implications of alternative allocation are substantial (see Manne and Richels, 1996; Richels et al, 1996). Results of studies for developing countries are even more striking. For instance, the 'grandfathered' and 'equal per capita' entitlement schemes for India lead to substantial losses or gains respectively, amounting to several per cent of India's GDP (see Shukla, 1996a; Fisher-Vanden et al, 1997). Equity, therefore, is vital and yet not trivial to achieve.

The present climate negotiations, culminating into the Kyoto Protocol, are restricted to emissions mitigation. Equity, therefore, is viewed within a narrow context of emissions entitlements, the distribution of which is an indirect equity criterion. An appropriate direct criterion is the minimization of global welfare losses resulting from mitigation and adaptation actions and impacts. The per capita income and emissions in developing countries, under a business-as-usual scenario, are expected to remain substantially below those in developed countries throughout the next century. A welfare-loss minimizing mitigation regime is therefore unlikely to allocate a substantial climate mitigation burden to developing countries. The alternative equity proposals, such as the convergence framework or the income-based entry, rest on indirect equity criteria. Ultimately, all such proposals will need to be endorsed by the three touchstones: justice, equity and efficiency. An efficient climate-change mitigation regime will require an early entry of developing countries. This will happen only if the post-Kyoto phase of climate negotiations quickly moves towards addressing the justice and fairness issues.

REFERENCES

Agarwal, A and Narain, S, 1991, *Global Warming in An Unequal World*, Centre for Science and Environment, New Delhi

Alcamo, J, Bouman, A, Edmonds J, Grubler, A, Morita, T and Sugandhy, A, 1995, Evaluation of the IPCC IS92 Emission Scenarios, *Climate Change 1994: Radiative Forcing of Climate Change and An Evaluation of the IPCC IS92 Emissions Scenarios*, Cambridge University Press, Cambridge, UK

Banuri, T, Goran-Maler, K, Grubb, M, Jacobson, H K and Yamin, F, 1996, Equity and social considerations, in: J P Bruce, H Lee and E Haites, eds, *Climate Change 1995: Economic and Social Dimensions of Climate Change*, Cambridge University Press, Cambridge, UK

Benestad, O, 1994, Energy needs and CO_2 emissions: Constructing a formula for just distributions, *Energy Policy* 22(9):725–734

Bodansky, D, 1993, The UN Framework Convention on Climate Change: A commentary, *Yale Journal of International Law* (Summer) 18:451–458

Chichilnisky, G and Heal, G, 1993, Global environmental risks, *Journal of Economic Perspectives*, 7(4):5–86

Chichilnisky, G and Heal, G, 1994, Who should abate carbon emissions? An international viewpoint, *Economic Letters* 44:443–449

Coase, R, 1960, The problem of social cost, *Journal of Law and Economics* 3:1–44

Edmonds, J and Wise, M, 1997, Exploring a technology strategy for stabilizing atmospheric CO_2, paper presented at the International Workshop on *Economic Impacts of Annex I Actions on All Countries*, Oslo, August 18–20

Flexner, S B, ed, 1987, *The Random House Dictionary of the English Language*, 2nd ed (unabridged) Random House, New York

Fisher-Vanden, K, Shukla, P R, Edmonds, J A, Kim, S H and Pitcher, H M, 1997, Carbon taxes and India, *Energy Economics* 19: 289–325

Grubb, M J, 1989, *The Greenhouse Effect: Negotiating targets*, Royal Institute of International Affairs, London

Grübler, A and Nakicenovic, N, 1991, International burden sharing in greenhouse gas reduction, Environmental Policy Division, World Bank, Washington, DC

Hayes, P, 1993, North–South transfer, in: P Hayes and K R Smith, eds, *The Global Greenhouse Regime: Who pays?*, pp 144–168, Earthscan Ltd, London

Houghton, J T, Meira Filho, L G, Callander, B A, Harris, N, Kattenberg, A and Maskell, K, eds, 1996, *Climate Change 1995: The Science of Climate Change, Contribution of Working Group I to the Second Assessment Report of the Intergovernmental Panel on Climate Change*, Intergovernmental Panel on Climate Change, Cambridge University Press, Cambridge

Hyder, T O, 1992, Climate Negotiations: The North/South Perspective, in: I Mintzer, ed, *Confronting Climate Change: Risks, Implications and Responses*, Cambridge University Press, Cambridge

Jepma, C J and Munasinghe, M, 1998, *Climate Change Policy: Facts, Issues and Analyses*, Cambridge University Press, Cambridge

Kverndokk, S, 1995, Tradable CO_2 emission permits: Initial distribution as a justice problem, *Environmental Values* 4:129–148

Kuznet, S, 1955, Economic growth and income inequality, *American Economic Review* 45(1):1–28

Leggett, J, Pepper, W J, Swart, R J, Edmonds, J, Meira Filho, L G, Mintzer, I, Wang, M X and Wasson, J, 1992, Emissions scenarios for the IPCC: An Update, in: *Climate Change 1992: The Supplementary Report to the IPCC Scientific Assessment*, Cambridge University Press, Cambridge, UK

Manne, A and Richels, R, 1996, The Berlin Mandate: The costs of meeting post-2000 targets and timetables, *Energy Policy* **24**(3):205–210

Pepper, W, Leggett, J, Swart, R, Wasson, J, Edmonds, J and Mintzer, I, 1992, Emissions scenarios for the IPCC, An Update, Assumptions, Methodology, and Results, Report Prepared for the Intergovernmental Panel on Climate Change, Working Group I

Rawls, J, 1971, *A Theory of Justice*, Oxford University Press, Oxford

Rayner, S and Malone, E, eds, 1998, *Human Choice and Climate Change: An International Assessment*, Battelle Press, Columbus

Richels, R, Edmonds, J, Gruenspecht, H and Wigley, T, 1996, *The Berlin Mandate: The design of cost-effective mitigation strategies*, Draft Paper, presented in IEA Conference on Climate Change Strategies, May 30–31, Paris

Rose, A, 1990, Reducing conflict in global warming policy: Equity as a unifying principle, *Energy Policy* **18**:927–935

Rose, A and Stevens, B, 1993, The efficiency and equity of marketable permits for CO_2 emissions, *Resources and Energy Economics* **15**:117–146

Schmiel, D, Enting, I G, Heimann, M, Wigley, T M L, Raynaud, D, Alves, D and Siegenthaler, U, 1995, CO_2 and the Carbon Cycle, *Climate Change 1994: Radiative Forcing of Climate Change and An Evaluation of the IPCC IS92 Emissions Scenarios*, Cambridge University Press, Cambridge, UK

Shukla, P R, 1996a, The modeling of policy options for greenhouse gas mitigation in India, *AMBIO* XXV(4)

Shukla, P R, 1996b, When and where aspects of climate change mitigation actions: A developing country perspective, Proceedings of IPCC Symposium on 'Integrated Assessment Process', Toulouse, France, October 24–26

Smith, K R, Swisher, J and Ahuja, D R, 1993, Who pays (to solve the problem and how much)?, in: P Hayes and K R Smith, eds, *The Global Greenhouse Regime: Who pays?*, pp 70–98, Earthscan Ltd, London

Toman, M and Burtraw, D, 1991, Resolving equity issues: Greenhouse gas negotiations, *Resources* **103**:10–13

UNFCCC (United Nations Framework Convention on Climate Change), 1992, Articles, United Nations, New York

UNFCCC (United Nations Framework Convention on Climate Change), 1997, Kyoto Protocol to the United Nations Framework Convention on Climate Change, United Nations, New York

Wigley, T, Edmonds, J and Richels, R, 1996, Economic and environmental choices in the stabilization of atmospheric CO_2 concentrations, *Nature* **379**:240–243

Wigley, T M L, Jain, A K, Joos, F, Nyenzi, S and Shukla, P R, 1997, Implications of proposed CO_2 emissions limitations, Intergovernmental Panel on Climate Change, Geneva

10 JUSTICE IN THE GREENHOUSE: PERSPECTIVES FROM INTERNATIONAL LAW

Frank Biermann[1]

INTRODUCTION

It may sound odd, but in practice, justice or fairness are hardly matters of concern for international lawyers. The rule that 'lawyers *find* the law and do not *make* it' could be seen as a basic tenet of all modern schools of legal thought. This necessarily results in a passive, apolitical attitude towards the notion of justice. For lawyers, justice is done when the law of the land or the law of nations is applied. The material content of this positive law will not be questioned, except for typical juridical investigations in the consistency of the normative system. Lawyers 'find the law' in the acts of the law-making bodies, whether they are governments in the national realm, or states and international organizations in the international sphere.

How can lawyers then contribute to the interdisciplinary debate on equitable and just solutions to the climate problem or other global environmental issues? Are there, for instance, political solutions to the climate problem that are closer to a concept of justice than others – from the point of view of international law? How can the scientific modelling and integrated assessment of global change profit from international legal science?

These questions will be addressed in this chapter. The argument is that we can derive general principles of justice from regularities in positive international environmental law that are considered to be just by the majority of governments. To this end, this chapter will first discuss the relationship between empirical state practice and normative concepts of justice in international law. Secondly, it will outline some elements of a theory of justice vis-à-vis global environmental problems.

---

[1] Many thanks to Carsten Helm, Leo-Felix Lee, Ferenc Tóth, Udo E Simonis and two anonymous referees for valuable comments on earlier versions of this chapter. All views expressed are the author's.

JUSTICE DERIVES FROM STATE PRACTICE

The Statute of the International Court of Justice states in Article 38, paragraph 1, that three sources of law are to be applied by the Court: international conventions (treaties) that establish rules expressly recognized by the contesting states; international custom, as evidence of a general practice accepted as law; and general principles of law. To put it in a rather polemical fashion, the Court of Justice will not explicitly rely upon justice. Instead, an international order is conceived as just when international conventions, international custom and general principles of law are strictly obeyed – no matter what their content.

Treaties are voluntarily concluded written agreements between states (Vienna Convention on the Law of Treaties, 1969). In this respect, states may agree on whatever they choose; this is part and parcel of their very sovereignty as states. There are, however, a few fundamental restrictions placed upon states' discretion by the modern concept of *ius cogens* (compulsory law), which evolved in the last decades in view of the crimes against humanity committed by Germany and Japan during World War II.[2] According to the concept of *ius cogens*, states must not conclude treaties that support wars of aggression, genocide, slavery, apartheid, or similar outrageous acts. But states may well conclude treaties that other states, or even they themselves, consider to be unjust.

The same applies to international customary law, the second source of international legal obligations. Customary law contains an objective element – that is, states must consistently conform to a common practice for some time – and a subjective element – that is, states must consider this common practice as the implementation of international legal duties and not as some form of courtesy or random uniform behaviour. In legal theory, we thus have a set of legal norms, derived from written or customary law, and a high degree of state compliance with these norms will be considered as 'justice that is being done'.

The third source of international law is 'the general principles of law recognized by civilized nations'. At first, one might wonder which nations will qualify here due to their state of civilization: today, all writers agree that no such distinctions are legally possible.[3] In some cases, the reliance upon general principles of law may open the door to include considerations of justice in international law (Verdross and Simma, 1984:380–394). For example, the International Court of Justice ruled in 1982 in the Continental Shelf case (*Tunisia versus Libya*) that in those (rare) cases where the court can choose among several possible interpretations of positive law, it is bound to select the interpretation that appears, in the light of the circumstances of the case, 'to be

---

2 *Ius cogens* has been codified in Article 53 of the Vienna Convention on the Law of Treaties (1969), and affirms that a treaty is void if it conflicts with a 'peremptory norm of general international law' – that is, norms that are 'accepted and recognized by the international community of states as a whole as a norm from which no derogation is permitted and which can be modified only by a subsequent norm of general international law having the same character'. Accordingly, Article 64 reads: 'If a new peremptory norm of international law emerges, any existing treaty which is in conflict with that norm becomes void and terminates.'

3 The phrase stems from the end of World War I, when former German and Italian colonies were not let into independence but subjected to the trusteeship of 'more civilized nations', such as South Africa.

the closest to the requirements of justice' (ICJ, 1982: paragraph 71; see, on this case, Yamin, 1994).

In practice, however, this works only *within* the legal relationship in a certain dispute (*intra leges*), not without (ICJ, 1969; Beck, 1994:203–204; also United Nations Department for Policy Coordination and Sustainable Development (UN DPCSD), 1995: paragraph 39). Firstly, the court will consider treaties and international custom, and only when two interpretations are possible, justice enters the legal analysis as part of the general principles of law recognized by civilized nations (these principles are subsidiary in fact, though not in doctrine). Such cases are rare; they mainly deal with border delimitation on the continental shelf.

Few legal instruments – such as the United Nations Convention on the Law of the Sea (UNCLOS) – explicitly call upon states to find 'equitable' or 'just' solutions vis-à-vis certain conflicts, such as the delimitation of fisheries zones (for example, UNCLOS, 1982: Article 59, 74, paragraph; Article 1, 82, 83, paragraph 1; and Article 140, paragraph 2). In a similar fashion, Article 3, paragraph 1 of the Climate Convention (UNFCCC, 1992) mentions the need for 'equity'. Also, some provisions of the biodiversity convention (UNCBD, 1992) refer to the 'equitable sharing of benefits' that may arise out of the utilization of genetic resources (Harris, 1997; UN DPCSD, 1995: paragraph 40). And yet, what equity eventually means in detail is left for states to decide: it is still 'negotiated justice'. For example, it is hard to imagine that any decision supported by the majority of parties to the Climate Convention could ever be challenged by legal action before the World Court only on grounds that it violates 'the principle of equity' enshrined in Article 3, paragraph 1, of that convention.

Consequently, in actual state behaviour many international legal rules emerged that some parts of the community of states apparently do not regard as just, but that are binding even upon the complaining parties. For example, the Antarctica regime, based on the Antarctic Treaty (1959), is seen by many developing countries as unjust because it privileges its original members and places costly obligations upon third parties that wish to join the treaty. The developing countries thus demanded, eventually with no success, that the regime should be changed and Antarctica should be set under UN rule as a 'common heritage of humankind'. An example from the Northern perspective is the rule that every nation – no matter its size or wealth – should have equal voting rights in international diplomatic conferences, which is part of the legal bedrock of the United Nations Organization. This is seen as unjust by the richer countries of the North which pay the lion's share of UN expenses without having a lion's share of the votes.

These examples show that the law of nations includes binding obligations which some nations consider unjust. The examples also indicate that different perceptions of justice often follow the North–South cleavage.

The first conclusion, therefore, is that justice does not directly follow from the doctrines of international law. Essentially, there is no *universal* concept of justice in international legal science superceding the positive law enacted by governments. The predominant notion of justice in legal science is the concept of 'formal justice', that is – in the words of John Rawls – justice as 'the impartial and consistent administration of laws and institutions, whatever their substantive principles' (Rawls, 1971: paragraph 10; in paragraph 38, Rawls speaks of 'justice as regularity' as a 'more suggestive phrase' than 'formal justice').

JUSTICE IN INTERNATIONAL ENVIRONMENTAL POLICY: THREE EMERGING PRINCIPLES

If there are no abstract concepts of justice in international law, how can international lawyers then approach justice? Their starting point will be the same as for every analysis of international law: actual state practice, together with the perception of states in the issue area under investigation. When we look at the perceptions of states and their actions and examine which legal norms states view as just, then the ozone layer protection regime strikes us as a prominent example (Vienna Convention, 1985; Montreal Protocol, 1987, amended in 1990 and 1992; see Benedick, 1998). Governments repeatedly stated that this regime has been effective because of the 'partnership, with common but differentiated responsibilities, that it had established between developing and developed countries' (UNEP/OzL.Pro.7/12, 1995: paragraph 59). Apparently, they see this partnership 'at the heart of the ozone regime', which is 'based on a balanced and equitable sharing of efforts and commitments' between North and South (UNEP/OzL.Pro.7/12, 1995: paragraph 71). Among all global environmental problems threatening human survival, governments view the ozone regime as 'more than treaties in and of themselves [but...] as models of co-operation on a global scale' (UNEP/OzL.Pro.7/12, 1995: paragraph 67).

To this author's knowledge, there are today no governments that consider the amended Montreal Protocol as an unjust treaty. The Montreal Protocol appears as a perfect example for just and equitable international environmental agreements, undisputed by all parts of the international community.

From this finding, international lawyers can approach the *general* question of justice in international environmental policy. To this end, this chapter will detail the following perspective: since governments declared that the Montreal Protocol was more than a treaty in that it was a model of just cooperation on a global scale, individuals should look at other environmental issues and ask whether the model of the Montreal Protocol had been followed in actual state practice. By so doing, general principles of justice in international environmental law can be derived that could also be applied to the question of justice in the future climate policy. This chapter will outline some emerging legal principles which could be regarded as the essence of a general concept of justice in international environmental policy.

More capable states shall accept more duties

Most scientific models on the impacts of climate change need to include assumptions on the future distribution of climate reduction measures across nations. International lawyers cannot hypothesize how reduction duties will be distributed, but they can analyse which distribution curves could be regarded as closer to a just solution according to the emerging principles of present international environmental law.

If we take the Montreal Protocol as a just agreement, we find that duties have been differentiated between two groups of countries: firstly, developing countries who consume per capita less than 300 grammes of CFC and who may delay their reduction timetables by ten years, including some other privileges; and secondly, all other countries. Regarding the definition of the term developing countries, Northern negotiators accepted the principle of self-definition – that is, all members of the political Group of 77 were privileged as developing countries without questioning the composition of that group (UNEP executive director, 1989: paragraphs 60–63; UNEP/OzL.Pro.1/5, 1989: Dec. I/12/E). It is important to note that consumption and production thresholds in the ozone regime – which defines privileged developing countries – have been based purely on per capita indicators, which is again argued for by Southern negotiators, especially from Africa, in climate negotiations. The differentiation criteria in the ozone regime eventually had the effect that only very few countries did not qualify for developing country privileges for some time, such as Cyprus and Slovenia or wealthy oil countries such as Kuwait (Victor, 1996).

A similar differentiation has already been incorporated within the climate convention. All members of the Group of 77 were granted special status. Whereas industrialized countries from West and East have to 'limit' their emissions, developing countries only have to develop policies to 'address their emissions', without any duty regarding the effect of those policies (UNFCCC, 1992: Article 4, paragraphs 1–2). Unlike the ozone treaty, the Group of 77 has not been further qualified according to criteria such as individual responsibility or capability. Thus, countries such as South Korea or Kuwait enjoy a more favourable status under the climate treaty than under the ozone treaty. Despite some resistance by the US, Australia, Canada and a few other industrialized countries, the Kyoto Protocol (1997) to the Climate Convention has maintained this basic differentiation between North and South.

All reduction obligations under the Kyoto Protocol thus apply only to Northern countries. These reductions duties (until the period 2008 to 2012) have been differentiated among industrialized countries, ranging from reduction duties of 8 per cent (for example, in the EU) to the right to increase emissions up to 10 per cent (Iceland). The legal assessment of this North–North differentiation is somehow difficult, since parties could not agree on any clear formula for their differentiation of obligations. As in the case of Russia and the Ukraine, relative capabilities may have played a role, as well as high percentages of renewable energy consumption (such as in Iceland). But it is too early to derive legal principles from these criteria, the more so since the Kyoto outcome is better explained by analysing actual bargaining power than by any moral or legal conceptual framework.

In the Biodiversity Convention, the principle of differentiation has been spelled out less explicitly, and detailed lists of countries have not yet been included within the regime. However, the preamble states that 'economic and social development and poverty eradication are the first and overriding priorities of developing countries', and several treaty provisions oblige the Conference of the Parties to take the special needs of developing countries into account. Most provisions also differentiate between parties by inserting phrases such as 'in accordance with its particular conditions and capabilities' or 'as far as possible and as appropriate' (see UNCBD, 1992: Article 6, paragraph 1, Article 7, paragraph 1, Article 8, paragraph 1 *et passim*).

Although the empirical evidence is still limited to a few cases, its consistency could lead to the conclusion that these provisions indicate the increasing acceptance of a general principle of differentiation; in other words, in global environmental regimes the different responsibilities and different capabilities of nations must be taken into account.

The recognition of different *responsibilities* follows from the general concept of international law – and, one could argue, of basic morality: that '[e]very international wrongful act of a state entails the international responsibility of that state' (ILC, 1980: Article 1). Whether there are 'wrongful acts' in the case of the greenhouse effect is certainly open to debate. The Climate Convention itself cannot be applied for past activities and, having permitted citizens to burn fossil fuels in past decades will hardly provide evidence for intentional wrongful acts by industrialized countries. Moreover, the essential causal link and the significance of any transboundary pollution will be difficult to prove if we analyse questions such as the responsibility of, say, Germany versus island states that may be inundated in the future. If Germany emits a certain percentage of global carbon dioxide emissions, will it be held 'responsible' for the inundation of the respective proportion of the Maldives Islands?

On the other hand, it is hardly acceptable if poor low-lying island states are flooded without any other state being held responsible. Principle 21 of the Stockholm Declaration on the Human Environment (1972) states that 'States have the responsibility to ensure that activities within their jurisdiction or control do not cause damage to the environment of other States'. In practice, the question of responsibility has often been mentioned in negotiations on the ozone issue, and responsibility has explicitly been included in the Climate Convention as justification for the differentiation of legal duties between industrialized and developing countries.

The consideration of different *capabilities* means that to solve collective problems, less capable nations should not be obliged to do as much as more capable nations. Again, the legal basis for a differentiation based on the capabilities of states is difficult to trace. The basis could be found in the internationally accepted 'right to development' that has been declared by the United Nations General Assembly in 1986 and is now universally recognized (after the last objector, the US, ceded). Similarly, widely accepted social human rights may entail the notion that the basic means for human survival and a decent living must not be taken from the individual. From both the social human rights and the 'collective' right to development, it follows that the means necessary to pursue these rights must not be taken from the individual person or the individual nation as long as conflicting rights are not involved (such as the rights of others with comparable significance). This leads to the conclusion that climate protection measures must not result in the impoverishment of such nations.

Taken together, when the less capable and the less responsible nation should do less to solve problems of the global community, it is quite sensible that in concrete environmental regimes, this nation will have to accept less stringent restrictions on environmentally harmful activities. This is the emerging 'principle of differentiation'. With regard to the integrated assessment of global change, this principle could also be seen as a prerequisite for any climate change modelling exercise.

More capable states shall assist others

Less reduction obligations for less capable and less responsible nations eventually result in less protection of the global environment and thus in its destruction. In this case, the rights of present generations are protected, but not the rights of future generations (Brown Weiss, 1997; UN DPCSD, 1995: paragraphs 41–47). This requires that in the medium term *all* nations – including the less capable and less responsible ones – must adopt effective environmental policies. Here, justice among nations requires the more capable and more responsible nations to compensate their neighbours for environmental protection programmes. This could be regarded as the 'principle of environmental solidarity': it is explicitly incorporated in the Montreal Protocol (1990: Article 10, paragraph 1), the Climate Convention (UNFCCC, 1992: Article 4, paragraph 3) and the Biodiversity Convention (UNCBD, 1992: Article 20, paragraph 2).

According to these treaties, industrialized states must provide new and additional resources to developing countries in order to meet their full agreed incremental costs. The agreement on the definition of 'incremental costs' rests in each case on the Conference of the Parties to the treaties and their subsidiary bodies. All treaties state that the fulfilment of developing countries' environmental duties depends on the fulfilment of the solidarity duties of industrialized countries. The Montreal Protocol, for example, provides in Article 5, paragraph 5 that assisting developing parties to comply with the control measures 'will depend upon the effective implementation of the financial co-operation' by industrialized countries as stipulated elsewhere in the treaty. This provision has been reiterated, *mutatis mutandis*, in Article 20, paragraph 4, of the Biodiversity Convention and Article 4, paragraph 7, of the Climate Convention.

Until now, only the refundable incremental costs of ozone-layer protection measures in developing countries have fully been defined. In order to implement this financial mechanism under the ozone regime, an independent fund has been set up in Montreal, which is governed by an executive committee with seven industrialized countries and seven developing countries with equal voting rights. The ozone case might indicate that governments are willing to interpret the term 'incremental costs' broadly (Biermann, 1997). The 'indicative list of categories of incremental costs', agreed upon in 1990, included, among others, cost categories such as the capital costs of converting existing production facilities; of establishing new production facilities for substitutes with capacities equivalent to the former facilities; the costs of premature retirement or enforced idleness of existing production facilities or the premature replacement of user equipment; all costs of patents, designs and the incremental cost of royalties; the costs of retraining workers; and the costs of research to adapt new technologies to local circumstances and to develop alternatives when ozone-depleting substances are used in the production as intermediate goods (UNEP/OzL.Pro.2/3, 1990: Appendix I of Annex IV to Dec II/8). Moreover, the ozone fund's executive committee must respect the industrial strategy of individual developing countries and avoid deindustrialization and the loss of export revenues in the South. The committee may also add any items to the list (on the implementation, see Biermann, 1997).

Rarely before have developing countries been granted such explicit legal rights on the transfer of money and technology from the North. However, the total costs of the ozone fund – slightly more than half a billion US dollars – are less than those required by similar future programmes on climate change and biological diversity. This could result in different definitions of the incremental costs within those regimes, when developing countries begin to effectively implement these conventions, and it could also result in conflicting views on justice in global climate and biodiversity policies.

Equal participation shall be guaranteed and participation shall be transparent

An important notion of justice is procedural justice, although this is not the main issue discussed in this book. Under present international law, the various modes of decision-making could be divided into two broad, ideal-type groups: either the decision-making follows the principle of one country, one vote, or the voting rights are weighted according to the 'importance' of the state in the respective issue area. The first alternative is conceived as unjust by industrialized countries, and it is mainly used in deliberative institutions with no effective decision-making powers, such as in the UN General Assembly. Developing countries, on the other hand, view weighted votes as unjust and undemocratic, because such procedures privilege, in most cases, industrialized countries, for example in the World Bank, the International Monetary Fund or the UN Security Council (see the criticism raised in South Centre, 1996).

The new treaties on international environmental policy might indicate a different, third solution to international decision-making – and this appears to be conceived as just by all parts of the international community. This is the emerging 'principle of equal participation in decisions'. It has been fully elaborated upon in the 1990 amendment to the Montreal Protocol. Here, decisions of the meeting of the parties require a two-thirds majority that must include the simple majority of the industrialized countries and at the same time the simple majority of the developing countries. North and South thus gained effective veto rights.

Comparable decision-making procedures were adopted for the executive committee of the ozone fund and, in 1994, for the reformed Global Environment Facility of the World Bank, administered jointly with the United Nations Environment Programme (UNEP) and the United Nations Development Programme (UNDP) (GEF, 1994). The GEF will finance global programmes on four issue areas: climate change, biodiversity, international waters and ozone depletion.[4] The reformed GEF is particularly interesting. Whereas decision-making procedures in the World Bank itself favour the North due to the bank's one-dollar, one-vote modus, in the bank's Global Environment Facility developing countries may now block any decisions. This is not the new international economic order as proclaimed by the UN General Assembly in the 1970s – but it is certainly more than industrialized countries are prepared to grant in other issue areas.

---

4 The GEF supports programmes on ozone-depleting substances only in countries that do not qualify for support by the fund under the Montreal Protocol. These are less wealthy industrialized countries (mainly in Eastern Europe) or developing countries that surpass the pertinent thresholds.

'COMMON CONCERN OF HUMANKIND' AS AN EMERGING CONCEPT OF INTERNATIONAL ENVIRONMENTAL LAW

In 1988 and 1992 respectively, the protection of the climate system and the preservation of the Earth's biological diversity were declared by the international community as a 'common concern of humankind'.[5,6] This legal notion is new. It evolved around the need for states to define their relationship as a community towards such global issues that affect all and that theoretically must imply a certain restriction of state sovereignty. When the declaration of common concerns of humankind was discussed in the United Nations General Assembly and intergovernmental negotiation committees, several other competing concepts were suggested by different states and academic writers. Some of them eventually appeared to be unfeasible, such as the attempt to directly extend the human rights regime to global environmental threats (Pathak, 1992) or the attempt to rely upon concepts of 'environmental security' (Timoshenko, 1992). Other concepts were rejected because some parts of the international community viewed them as unjust, in particular the notion of a 'common heritage of humankind' that some states had suggested as a basic legal concept for global climate and global biodiversity.[7]

---

5 See UNGA, 1988. The Assembly Resolution reads: *The General Assembly, welcoming with appreciation* the initiative taken by the Government of Malta by proposing for consideration by the Assembly the item entitled 'Conservation of climate as part of the common heritage of mankind',
   Concerned that certain human activities could change global climate patterns, threatening present and future generations with potentially severe economic and social consequences,
   Noting with concern that the emerging evidence indicates that continued growth in atmospheric concentrations of 'greenhouse' gases could produce global warming with an eventual rise in sea levels, the effects of which could be disastrous for mankind if timely steps are not taken at all levels...
   (1) *Recognizes* that climate change is a common concern of mankind, since climate is an essential condition which sustains life on earth;
   (2) *Determines* that necessary and timely action should be taken to deal with climate change within a global framework...
   (6) *Urges* governments, intergovernmental and non-governmental organizations and scientific institutions to treat climate change as a priority issue, to undertake and promote specific, cooperative action-oriented programmes and research so as to increase understanding on all sources and causes of climate change, including its regional aspects and specific timeframes as well as the cause and effect relationship of human activities and climate, and to contribute, as appropriate, with human and financial resources to efforts to protect the global climate...
   (9) *Calls upon* governments and intergovernmental organizations to collaborate in making every effort to prevent detrimental effects on climate and activities which affect the ecological balance, and also calls upon nongovernmental organizations, industry and other productive sectors to play their due role...

6 See the Climate Convention (UNFCCC, 1992) and the Biodiversity Convention (UNCBD, 1992) – in both cases the preamble.

7 In the 1970s, the developing world had battled for the notion of a common heritage of humankind for exploitable international commons such as manganese ores from the deep sea and the resources of the moon and Antarctica. In the late 1980s, when the protection of global environmental systems entered the diplomats' stage, it was industrialized countries who supported the notion of a common human heritage, and the developing countries who rejected the idea because they feared to lose their sovereignty. The notion of a common supranational heritage, when used for activities inside the national territories, was seen by the South as heralding new environmental colonialism. It is a widely held misunderstanding by environmental non-governmental organizations and some academic writers that the common concern approach is somehow less than the common heritage concept, and that the first is only watering down the second (on different views, see Schrijver, 1996). To save our planet's ecosystems from human activities, the common concern approach could be as effective as the common heritage approach.

Instead, all states supported the concept of 'common concern of humankind' with regard to the dangers of climate change and the depletion of biodiversity. It appears reasonable to extend this notion 'backwards' to the older treaties on ozone depletion, since these agreements – addressing essential global problems – had already been concluded before the concept of common concern of humankind was referred to for the first time.

It is difficult to pinpoint the substance of the concept of common concern of humankind. What does it mean for lawyers and political practitioners; which explicit norms may follow from this concept? Elsewhere I have elaborated upon this question in more detail (Biermann, 1996). With regard to the question of justice, I would argue that the three emerging principles suggested in this chapter could be seen as elements of a theory of justice in international environmental affairs, as well as elements of the general legal concept of common concern of humankind.

However, the concept of common concern of humankind includes a final and very important principle which does not, nevertheless, take precedence in this book on justice and equity (see Figure 10.1). In tandem with the emerging principles of differentiation, solidarity and equal participation, I would argue that individual nations may no longer rely upon their sovereignty when most states consider environmental problems as a common concern of humankind that requires effective environmental policies on a global scale. Technically speaking, this would imply modifying the doctrine of the sources of international law. While traditionally only those states have been bound by customary law which have at least implicitly consented to the legal norm in question, they may now be forced to obey, even without their consent. A strong indication that states are accepting such issue-specific limitations for state sovereignty is Article 4 of the Montreal Protocol, which provides for a complex trade restriction regime vis-à-vis third parties who refused to join the treaty.

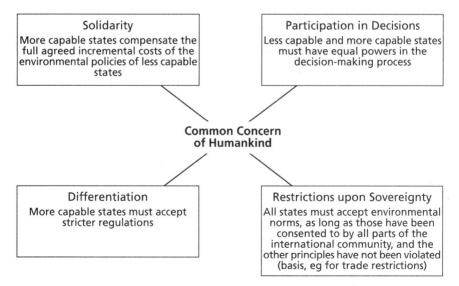

Figure 10.1 The norm square of the emerging legal concept of common concerns of humankind

This leads to a final point: the practical consequences of legal theory. The practical effect of legal theorizing could be, for example, that collective measures against free-riders could be made more legitimate. Public pressure could be expanded and trade-restrictive measures would be justified, in certain cases even when trade restrictions were in breach of the General Agreement on Tariffs and Trade (GATT). This legal framework – in technical terms the law of retorsion and reprisal – would set the current debate on trade and environment (Helm, 1996) on a better and also fairer conceptual footing than the reliance upon the notion of ex-territorial application of internal environmental law. The yardstick of legitimized trade restrictions against free-riders could be directly derived from the principles of differentiation, solidarity and equal participation (of North and South) in decision-making.

In any case, such restrictions of national sovereignty are acceptable and justified under international law only when a 'common concern of humankind' is at stake. But this again implies that the principles of differentiation, solidarity and equal participation must underlie any environmental regime that is to be enforced upon remaining minorities of the international community against their sovereign will. The concept of common concern of humankind is thus not a menu – it is a complex set of emerging principles that are interrelated and can only be applied as such. But empirical evidence suggests that these principles – if applied together – are judged as just by all parts of the international community. Those interrelated principles may thus be regarded as the bedrock of justice in international environmental law under the overall concept of common concern of humankind.

REFERENCES

Antarctic Treaty, Washington DC, 1 December 1959, in force 23 June 1961, in: 402 UNTS 71

Beck, H, 1994, *Die Differenzierung von Rechtspflichten in den Beziehungen zwischen Industrie- und Entwicklungsländern. Eine völkerrechtliche Untersuchung für die Bereiche des internationalen Wirtschafts-, Arbeits- und Umweltrechts*, Peter Lang Publishers, Frankfurt am Main, Germany

Benedick, R E, 1998, *Ozone Diplomacy: New Directions in Safeguarding the Planet*, Harvard University Press, Cambridge (Mass), US

Biermann, F, 1995, *Saving the Atmosphere: International Law, Developing Countries and Air Pollution*, Peter Lang Publishers, Frankfurt am Main, Germany

Biermann, F, 1996, Common concern of humankind: The emergence of a new concept of international environmental law, *Archiv des Völlkerrechts* 34(4):426–481

Biermann, F, 1997, Financing environmental policies in the South: Experiences from the multilateral ozone fund, *International Environmental Affairs* 9(3):179–218

Biermann, F, Büttner, S, and Helm, C, eds, 1997, *Zukunftsfähige Entwicklung: Herausforderungen an Wissenschaft und Politik, Festschrift for Udo E Simonis*, edition sigma, Berlin, Germany

Brown Weiss, E (ed), 1992, *Environmental Change and International Law*, United Nations University Press, Tokyo, Japan

Brown Weiss, E, 1997, Treating future generations fairly, in: F Biermann, S Büttner and C Helm, eds, *Zukunftsfähige Entwicklung: Herausforderungen an Wissenschaft und Politik, Festschrift for Udo E Simonis*, edition sigma, Berlin, Germany, 126–141

GEF (Global Environment Facility), 1994, Participants, instrument for the establishment of the restructured Global Environment Facility, 16 March 1994, in: 33 *International*

Legal Materials 1273 (1994)
Harris, P G, 1997, Environment, history and international justice, *The Journal of International Studies* (Tokyo) 40(July), 1–33
Helm, C, 1996, Transboundary environmental problems and new trade rules, *International Journal of Social Economics* 23(8):29–45
ICJ (International Court of Justice), 1969, *North Sea Continental Shelf* Cases (*Federal Republic of Germany versus Denmark; Federal Republic of Germany versus The Netherlands*), International Court of Justice, 1969 ICJ Report 3
ICJ (International Court of Justice), 1982, *Case Concerning the Continental Shelf* (*Tunisia versus Libyan Arab Jamahiriya*), 24 February 1982, 1982 ICJ Reports 60
ILC (International Law Commission), 1980, *Draft Articles on State Responsibility* 1980, in: ILC Yearbook II, Part II, 26
Kiss, A and Shelton, D, 1991, *International Environmental Law*, Transnational Publishers, Ardsley-on-Hudson
Kyoto Protocol, 1997, Kyoto Protocol to the United Nations Framework Convention on Climate Change, Kyoto, 10 December 1997, not in force, in: *Conference of the Parties to the United Nations Framework Convention on Climate Change*, Kyoto Protocol to the United Nations Framework Convention on Climate Change, adopted at its III Session (Kyoto, 1–10 December 1997), UN Doc FCCC/CP/1997/L 7/Add 1 of 10 December 1997
Montreal Protocol, 1987, Protocol (to the 1985 Vienna Convention on the Protection of the Ozone Layer) on Substances that Deplete the Ozone Layer, Montreal, 16 September, in force 1 January 1989, in: 26 International Legal Materials 1550 (1987)
Montreal Protocol, 1990, Amendment (and Adjustments) to the Montreal Protocol on Substances that Deplete the Ozone Layer, London, 29 June 1990, Amendment in force 10 August 1992, in: 30 *International Legal Materials* 537 (1991)
Montreal Protocol, 1992, Amendment (and Adjustments) to the Montreal Protocol on Substances that Deplete the Ozone Layer, Copenhagen, 23–25 November 1992, Amendment in force 14 June 1994, in: 23 *International Legal Materials* 874 (1993)
Pathak, R S, 1992, The human rights system as a conceptual framework for environmental law, in: E Brown Weiss, ed, *Environmental Change and International Law*, United Nations University Press, Tokyo, Japan, pp 205–243
Rawls, J, 1971, *A Theory of Justice*, The Belknap Press of Harvard University Press, Cambridge, Mass, US
Rio Declaration on Environment and Development, 1992 – Rio Declaration on Environment and Development, adopted at the United Nations Conference on Environment and Development (Rio de Janeiro, 3–14 June 1992), in: 31 *International Legal Materials* 874 (1992), at 876
Schrijver, N, 1996, Sovereignty: stumbling block or cornerstone of an active international environmental policy?, *Change* 32, September, 1–3
South Centre, 1996, *For a Strong and Democratic United Nations. A South Perspective on UN Reform*, South Centre, Geneva, Switzerland
Stockholm Declaration, 1972, Stockholm Declaration of the United Nations Conference on the Human Environment, 16 June 1972, in: 11 *International Legal Materials* 1416 (1972)
Timoshenko, A S, 1992, Ecological security: Response to global challenges, in: E Brown Weiss, ed, *Environmental Change and International Law*, United Nations University Press, Tokyo, Japan, pp 413–456
UNCBD, 1992, United Nations Convention on Biological Diversity, Rio de Janeiro, 5 June 1992, in force 29 December 1993, in: 31 *International Legal Materials* 818 (1992)
UNCLOS, 1982, United Nations Convention on the Law of the Sea, Montego Bay, 10 December 1982, in force 16 November 1994, in: 21 *International Legal Materials* 1261 (1982)

UN DPCSD (United Nations Department for Policy Coordination and Sustainable Development), 1995, *Report of the Expert Group Meeting on Identification of Principles of International Law for Sustainable Development, Geneva, Switzerland, 26–28 September 1995*, prepared by the Division for Sustainable Development for the Commission on Sustainable Development, Fourth Session, 18 April–3 May 1996, New York, US

UNEP Executive Director, 1989, *Report to the First Meeting of the Parties to the Montreal Protocol on Substances that Deplete the Ozone Layer* (Helsinki, 2–5 May 1989), UN Doc UNEP/OzL.Pro.1/2 of 28 March

UNEP/OzL.Pro.1/5, 1989, *Report of the First Meeting of the Parties to the Montreal Protocol on Substances that Deplete the Ozone Layer* (Helsinki, 2–5 May 1989), 6 May

UNEP/OzL.Pro.2/3, 1990, *Report of the Second Meeting of the Parties to the Montreal Protocol on Substances that Deplete the Ozone Layer* (London, 27–29 June 1990), 29 June

UNEP/OzL.Pro.7/12, 1995, *Report of the Seventh Meeting of the Parties to the Montreal Protocol on Substances that Deplete the Ozone Layer* (Vienna, 5–7 December 1995), 27 December

UNEP/OzL.Pro.ExCom/8/29/Corr 1, 1992, *Executive Committee of the Multilateral Fund for the Implementation of the Montreal Protocol on Substances that Deplete the Ozone Layer*, Report of the 8th Meeting, 12 November

UNFCCC, 1992, United Nations Framework Convention on Climate Change, New York, 9 May 1992, in force 21 March 1994, in: 31 *International Legal Materials* 849 (1992)

UNGA, 1988, United Nations General Assembly Resolution 43/53 (Protection of Global Climate for Present and Future Generations of Mankind) of 6 December 1988, in: 28 *International Legal Materials* 1326 (1989)

Verdross, A and Simma, B, 1984, *Universelles Völkerrecht Theorie und Praxis*, Duncker and Humblot, Berlin, Germany

Victor, D G, 1996, *The early operation and effectiveness of the Montreal Protocol's non-compliance procedure*, International Institute for Applied Systems Analysis, Laxenburg, Austria

Vienna Convention, 1985, Convention on the Protection of the Ozone Layer, Vienna, 22 March 1985, in force 22 September 1988, in: 26 *International Legal Materials* 1529 (1987)

Vienna Convention on the Law of Treaties, 1969, Convention on the Law of Treaties, Vienna, 23 May 1969, in force 27 January 1980, in: 8 *International Legal Materials* 679 (1969)

Yamin, F, 1994, Principles of equity in international environmental agreements with special reference to the climate change convention, paper presented at the IPCC workshop Equity and Social Considerations, Nairobi, Kenya

11 EQUITY IN INTERNATIONAL LAW

*Juliane Kokott**

INTRODUCTION

Recourse to concepts such as equity, fairness, justice or natural law becomes necessary in the absence of clear and precise rules of the positive law. This applies to the law of climate change.

The United Nations Framework Convention on Climate Change (UNFCCC, 1992), signed at the UN Conference on Environment and Development, was ratified by nearly all states of the world in an extraordinarily short period of time. It has contributed considerably to the increasing general awareness of problems relating to climate change. However, the other aspect of the fast and almost universal ratification of this convention is the fact that it contains no precise obligations for states – in particular, no precise obligations for industrialized countries to reduce the emission of detrimental greenhouse gases. This insufficiency of the convention will be partly remedied by the entry into force of the protocol adopted at the Kyoto conference in December 1997 (Kyoto Protocol, 1997). The most important aspect of this protocol is the determination of legally binding reduction rates for greenhouse gases. However, the protocol fixes reductions rates only for the first quantified emission limitation and reduction period from 2008 to 2012 and only for developed country parties and parties undergoing the process of transition to a market economy, mainly in Eastern Europe (Annex I parties).

The following chapter tries to clarify the legal principles or framework laid down in general public international law as well as in the convention itself, which could serve as guidelines in further concretizing the states' duties in the field of climate protection. The starting point, therefore, is Article 3 (1) of the convention according to which '[T]he Parties should protect the climate system for the benefit of present and future generations of humankind, on the basis of equity and in accordance with their common but differentiated responsibilities and respective capabilities.'

The convention, thus, seems to modify the classical international law principles of the formal equality of states in favour of a more 'equitable'

---

* Assisted by Dominik Thieme

approach, taking into account the differentiated responsibilities and respective capabilities, and the specific needs and special circumstances of particular states. Therefore, it seems useful to clarify the meaning of the term equity. This term has its roots in Aristotelian thought, was further developed by the English equity courts, where it had an affinity with 'natural justice' as opposed to 'strict law', and finally found its way into international law (Hendel, 1996). In international law, the term equity has undergone a real renaissance. The International Court of Justice (ICJ) increasingly has decided cases relating to the delimitation of maritime boundaries on the basis of equitable principles. Another increasingly important field of application of the concept of equity is modern environmental law, including the law on climate change.

The following sections show the development, various meanings and growing fields of the concept of equity in international law. On this basis, the last section examines which lessons this concept may provide us with in order to clarify the states' duties in the area of climate protection.

EQUITY IN GENERAL INTERNATIONAL LAW

The traditional perception of the concept of equity in international law is rooted in Aristotelian thought, where equity is defined as the corrective of law. Aristotle felt that the law-giver could draft laws in general terms only, and that consequently laws could not attain their intended purpose in every case. Accordingly, in special cases the law needs to be tempered by equity in order to achieve a just result – a result the law-giver would presumably have sought, if he had foreseen the exceptional case.

In the 19th and early 20th centuries, references to equity and the equitable powers of international arbitral tribunals appeared frequently. Following the conclusion of the peace treaties after World War I, there was considerable debate regarding the possible modification of those terms of the treaties which were unfair to Germany by courts given equitable competence to correct the legal rules.

These developments led to a doctrinal debate between adherents of the schools of positivism and natural law regarding the role of equity in international law. Because the concept of equity envisages that the judge will correct positive rules of law, positivists attacked its use as an unwarranted encroachment on state sovereignty and as an unwise grant of judicial discretion. Conversely, the advocates of natural law applauded equity and judicial discretion as a means of attaining more fairness and justice in international law.

Different functions of equity in traditional international law

In order to better understand the role of equity in international law, Akehurst, in his landmark article on equity and general principles of law, distinguished three ways in which an international judge might apply equity: equity within the law (*infra legem*), equity as a gap filler (*praeter legem*) and equity against the law (*contra legem*) (Akehurst, 1976; Janis, 1995).

Equity within the law (infra legem)

It is generally recognized that international courts and tribunals may apply equity within the law. In this sense, equity works within the limits of the law more or less as a rule of interpretation to ensure that the application of a rule results in an equitable – that is, fair or just – result. Also in this category of equity within the law fall cases where it is impossible to quantify the damages precisely. International courts and tribunals then make an equitable estimation of the compensation to which the claimant is entitled.

Equity as a gap filler (praeter legem)

Equity is used as a gap filler when international law is silent. The extent to which courts have used equity as a gap filler is rather limited. Some international decisions have claimed the power to fill gaps in the law by resort to equity; other international decisions deny that such a power exists. The reason for the caution of international courts is as follows.

International courts and tribunals depend on being accepted by the states which are under no obligation to submit cases to them. Thus, they rely on the voluntary cooperation of states. International courts fear that the decisions based on equity are not sufficiently predictable for states, with the consequence that states would submit even fewer cases to international courts and tribunals if they made an extensive use of equity (ICJ, 1984).[1]

Equity against the law (contra legem)

Equity against the law (*contra legem*) comes close to the traditional concept of equity in common law. If the general rules of the law are too strict as applied to a particular case, a decision based on the principles of equity may be more just.

New developments

In his monograph *Fairness in International Law and Institutions*, Tom Franck (1995) deals with equity as part of the broader concept of fairness. According to Franck, fairness is a composite of two independent variables: legitimacy and justice. He recommends studying the emerging role of equity in the jurisprudence of international tribunals in order to find out about the justice of international law (Franck, 1995:47). Equity in his sense implies elements of distributive justice as well as new and responsible approaches to common heritage goods (Franck, 1995:412).

Generally, equity in the context of modern international law is used not in the technical sense which the word possesses in Anglo-American legal systems in the distinction between law and equity as separate bodies of law, but as a synonym for 'justice'. Moreover, there are natural law connotations in modern equity discourse. Thus, the three terms – equity, justice and natural law – tend to merge into one another (Malanczuk, 1997).[2]

---

1 See p 385, dissenting opinion of Judge Gros.
2 For other recent studies on equity, see Rossi (1991) and Lachs (1993).

The International Court of Justice: distinguishing equity from decisions *ex aequo et bono*

Recourse to equity in traditional international law is usually distinguished from the power of courts to decide *ex aequo et bono*. This distinction is required since courts 'will always have regard to equity *infra legem*, that is, that form of equity which constitutes a method of interpretation of the law in force, and is one of its attributes' (ICJ, 1986:567). The power of courts to decide *praeter legem* or *contra legem*, to the contrary, is much more limited. In particular, in its decisions concerning the delimitation of maritime boundaries from 1969 onwards, the International Court of Justice (ICJ) made extensive use of equity or equitable principles, emphasizing that it did not decide *ex aequo et bono* (ICJ, 1969; 1982; 1984; 1985; 1986).[3] This emphasis was necessary because the Court is only allowed to decide *ex aequo et bono* with the special consent of the parties (Article 38 (2), Statute ICJ), which it did not have in those cases.

However, it is difficult to draw a clear distinction between decisions based on the legal principle of equity, or on equitable criteria, and decisions *ex aequo et bono*. One could translate the Latin term *ex aequo et bono* precisely by the English term equity in the way that both expressions are derived from the same Latin origin meaning 'fair and good'. The special feature of *ex aequo et bono* decisions is that the court does not even pretend to base its reasoning on legal principles and simply tries to reach a fair compromise with the consent of the parties (Janis, 1995). Some maintain that the ICJ did nothing else than determine maritime boundaries *ex aequo et bono* (ICJ, 1982; 1984; 1985).[4] But, in any case, the court could not do this openly, since the parties had not given it the special power to decide *ex aequo et bono*.

Summing up, one can conclude that equity in this more traditional context has been a problem mainly for the courts. It had, and still has, to do with the limits of the powers of international courts and tribunals.

An interesting phenomenon is the ever increasing importance of the principle of equity since the ICJ cases concerning maritime boundaries. In 1993, Judge Weeramantry of the ICJ wrote a 70-page separate opinion dealing exclusively with equity (ICJ, 1993). Judge Weeramantry correctly emphasized that equity is playing an increasingly important international role (ICJ, 1993:212, paragraph 118). He refers to 'concepts of a higher trust of Earth resources, an equitable use thereof which extends intertemporally, the "*sui generis*" status accorded to such planetary resources as land, lakes and rivers, the concept of wise stewardship thereof, and their conservation for the benefit of future generations' (ICJ, 1993:240, paragraph 235). The increased role of a modern concept of equity in international environmental law will be dealt with more extensively after delineating the role equity played in the debates about a new international economic order.

---

3 See ICJ (1969: 46); see also pp 132, 136 and 148 (separate opinion of Judge Ammoun); ICJ (1982: 59); see also p 290 (dissenting opinion of Judge Eversen); see also Art 83, para 1 of UNCLOS (1982); see material on terrestrial frontiers (ICJ, 1986:567).
4 ICJ, (1982: 157) (dissenting opinion of Judge Oda); ICJ (1984: 382) (dissenting opinion of Judge Gros); ICJ (1969: 166) (dissenting opinion of Judge Koretsky).

The new international economic order

The idea of a 'new international economic order' refers to equity outside the courtroom. The Charter on Economic Rights and Duties of States (UN Charter, 1974) and the Declaration of the General Assembly of the United Nations on a New International Economic Order (NIEO, 1974) aim at 'equitable' international economic relations and support for developing countries.

Moreover, discussions on a new international economic order have added the concept of distributive justice to the equity discourse. Not surprisingly, the idea of distributive justice as opposed to formal equality was particularly endorsed by the developing countries. These states see equity as a potential means to counterbalance their underdevelopment and poor economic conditions. The 1986 Seoul Declaration of the International Law Association (ILA, a private organization of international law specialists, which tries to codify international law) on the New International Economic Order (NIEO) also refers to the distinction between formal equality and true equality, achieved by applying the principle of equity.

> *Without ensuring the principle of equity there is no true equality of nations and states in the world community consisting of countries of different levels of development. A new international economic order should therefore be developed by the United Nations and international organizations, by treaties and by state practice in conformity with the principle of equity, which means that this development should aim at a just balance between converging and diverging interests and, in particular, between the interests of developed and developing countries.*
>
> ILA, 1986

Similarly, Judge M Bedjaoui, in his book on the new international economic order, emphasizes: 'This international law of participation ... must give great prominence to the principle of equity (which corrects inequalities) rather than to the principle of equality' (Bedjaoui, 1979:127). Thus, like the more recent Framework Convention on Climate Change, those supporting the New International Economic Order distinguish between the rights and obligations of developed and developing countries.

Developed countries, to the contrary, tend to rely more on the principle of the sovereign equality of states (Article 2 (1) of the Charter of the United Nations). Due to differences in perception between developing and developed countries, the idea of creating a 'new international economic order' has not won universal acceptance, at least not as a legal principle.

Nevertheless, equitable principles have become part of international economic law in various respects. For example, the General Agreement on Tariffs and Trade (GATT, 1947) and the Agreement Establishing the World Trade Organization (WTO, 1994; GATT, 1994) allow for a preferential treatment of developing countries and, thus, include within the mechanisms of the generalized system of preferences (GATT, 1979) elements of distributive justice (Langer, 1995:293).[5] This régime is designed to help developing countries

---

5 For the treatment of developing countries see WTO (1994: Preamble, para 2); GATT (1994: 1263, Declaration no 5 and p 1248, Decision on Measures in Favour of Least-Developed Countries, Preamble, para 3). Both the ministerial declaration and the decision recall GATT (1979).

increase their trade with developed countries in accordance with the 'equitable' principles laid down in the Charter of Economic Rights and Duties of States (UN Charter, 1974).[6]

Besides equity, documents and discourse relating to the New International Economic Order stress the principles of interdependence, participation, common interest and cooperation among all states.[7] These aspects are also essential elements of the principle of equity as applied in the expanding field of international environmental law.

Dangers of applying equity

Equity is subjective. The international community consists of states with very different cultural, ethical and political views, and it is doubtful whether a just decision for one party will be considered just and fair by other parties to the dispute or by other members of the international community as well. Natural law does not provide the necessary basis for objective equity either, since views of natural law vary from person to person and from century to century, just as views on equity do (Akehurst, 1976:812). Furthermore, the issues of disputes brought before the international tribunals are rather complex, and an equitable solution balancing the conflicting interests usually 'does not exactly leap to the eye' (Akehurst, 1976:810). Therefore, equity *contra legem*, in particular, should be applied with extreme vigilance.

Nevertheless, the development of an international public order within a 'global society' implies that there are more common values in modern international law than in traditional international law. These common values are best reflected in international human rights and international environmental law. Correspondingly, Akehurst, in his already mentioned landmark article on equity and general principles of law, points to the bold general principles-based equity approach of the Court of Justice of the European Communities in the field of human rights (Akehurst, 1976:810). Similarly, international environmental law implies more common values important to all states and the international community and leaves, therefore, more room for equity than other areas of public international law.

EQUITY IN INTERNATIONAL ENVIRONMENTAL LAW

In international environmental law, the principle of equity is both more important and more clearly defined than in general international law (Kiss and Shelton, 1991:121; Birnie and Boyle, 1992:126). This is connected with the ongoing transformation of international law: traditional international law primarily served the interests of the individual states and not a common or public interest. However, particularly since World War II, international law has more and more become an area of law which shows elements of national constitutional law insofar as more and more rules develop which do not only

---

6 Cf. Articles 18, 19 and 21.
7 See Preamble of the UN Charter (1974); Preamble of the NIEO (1974); see also the principle(s) as laid down in Article 4 (b).

serve the individual interests of particular states but also serve common interests or common concerns.[8] Those more recent rules of international law, particularly in the area of human rights and international environmental law, serve the interests of all states, the interests of mankind and even the interests of future generations (Riedel, 1997; Oxman, 1997).

This new international law also contains peremptory rules of international law (*ius cogens*), from which no derogation by treaty is permitted.[9] Furthermore, whenever human rights conventions, for example, prohibit derogations from certain rights even in cases of emergency, this suggests that we might deal with peremptory rules of international law. Obligations of essential importance for safeguarding human beings and for safeguarding and preserving the human environment, such as those prohibiting massive pollution of the atmosphere, are peremptory rules of international law. Infraction of these rules is not only an international delict, but an international crime.[10]

Another characteristic of modern international law are *erga omnes* obligations: obligations which a state owes to the community of states or to all states, and not only to a specific other state on the basis of treaty obligations.[11,12]

Both categories, *erga omnes* obligations and *ius cogens*, demonstrate that international law is more than a field which only concerns the state parties involved in a particular issue; rather, it deals more and more with common interests or common concerns. It is precisely in international environmental law or in conventions or treaties relating to shared resources, common heritage or common concern values that the principle of equity plays its most important role.

Multilateral environmental agreements and the concept of equity

Practically all multilateral environmental agreements dealing with shared resources, common heritage or common concern goods now contain the principle of equity.[13] Treaty provisions may refer to equity, equitable use or share, or cooperation and assistance in various ways.

---

8 On various approaches, see Brunnée (1989); Biermann (1996).
9 Article 53 of the Vienna Convention on the Law of Treaties (1969). See also Frowein (1984).
10 Cf ILC (1996: Part I, Art 19 (3) c and d).
11 The International Court of Justice defined obligations *erga omnes* as follows: 'an essential distinction should be drawn between the obligations of a State towards the international community as a whole, and those arising vis-à-vis another State in the field of diplomatic protection. By their very nature the former are the concern of all States. In view of the importance of the rights involved, all States can be held to have a legal interest in their protection; they are obligations *erga omnes*. Such obligations derive, for example, in contemporary international law, from the outlawing of acts of aggression, and of genocide, as also from the principles and rules concerning the basic rights of the human person, including protection from slavery and racial discrimination' (ICJ, 1970:33). See also Articles 40 f of ILC (1996).
12 On the relationship between obligations *erga omnes* and *ius cogens*, see Kimminich (1997); Kokott (1998a).
13 Apart from the treaties referred to in this chapter's second section, cf also UNEP (1978: Principles 1, 2 (Cooperation); 5–7 (Information)). See also Article 140 (2) of UNCLOS (1982).

The international protection of freshwater

International rules on the use of freshwater refer to various elements of equity. The Helsinki Rules of the International Law Association provide for a system of cooperation in order to guarantee an equitable use of this shared resource (ILA, 1966). The UN Watercourses Convention (1992) includes, among others, the following elements of equity in Article 2:[14]

1 The Parties shall take all appropriate measures to prevent, control and reduce any transboundary impact.
2 The parties shall, in particular, take all appropriate measures ... to ensure that transboundary waters are used in a reasonable and equitable way [...]
5 In taking the measures referred to in paragraphs 1 and 2 of this article, the parties shall be guided by the following principles: (a) The precautionary principle; (b) the polluter pays principle; (c) water resources shall be managed so that the needs of the present generation are met without compromising the ability of future generations to meet their own needs.

The recently adopted United Nations Convention on the Law of the Non-Navigational Uses of International Watercourses (UN Watercourses, 1997) points in the same direction: equitable use of a shared resource, including duties to cooperate and to inform each other.[15,16]

Duties to cooperate, notify and inform each other are closely related to principles of equity and fairness. This is so because equitable solutions may only be reached on the basis of taking into account and balancing the interests of those affected. These interests are best articulated and evaluated with the cooperation and participation of those affected by a particular decision or solution (Kokott, 1998b).

The older Rhinesalt Convention (1976) and its 1991 protocol choose a different approach. Instead of referring to the polluter pays principle and more general principles aiming at an equitable and fair burden-sharing, Article 4 of the convention distributes the costs for certain salt reduction measures – reduction measures undertaken on French territory up to a ceiling of 400 million French francs and reduction measures undertaken on Dutch territory up to a ceiling of 32.37 million Dutch guilders, according to fixed percentages among Germany (30 per cent), France (30 per cent), The Netherlands (34 per cent) and Switzerland (6 per cent). One of the reasons for this special regime probably is that The Netherlands, as the downstream country, is much more interested in reduction measures taking place in upstream France or Switzerland than vice versa. From this state interest perspective, it seems fair that those (downstream) states interested in and profiting from the reduction measures should contribute the lion's share. However, the increasing importance of protecting the environment, as well as *erga omnes* aspects of modern international environmental law, support a polluter pays or equity approach. This is due to the fact that the prevention of extensive river pollution is not

---

14 UN Watercourses (1992): Article 2 (General Provisions and Arts); Articles 5, 12 (Common Research and Development); Articles 6, 13 (Exchange of Information); Article 9 (Cooperation).
15 See also McCaffrey and Sinjela (1998); Kokott (1998b).
16 See UN Watercourses (1997) Articles 5, 6 and 8. On recent developments, see Brunnée and Toope (1997); Dellapenna (1994).

only in the interest of downstream state parties, but tends to become a common public interest of all states under modern international law.

United Nations Conference on Environment and Development (Rio 1992)

The main focus regarding the further development of international environmental law at the United Nations Conference on Environment and Development (UNCED) was on sustainable development. The principle of sustainable development entitles the present generation to a development sufficient to fulfill its needs without, at the same time, hindering future generations from fulfilling their future needs. The adjective sustainable, thus, modifies the right to development. Development has to take place in a way that does not destroy or use up non-replaceable resources of nature. This can be seen as an intergenerational or equitable approach to development.

The two conventions adopted at UNCED deal with the protection of biodiversity and the global climate. The Convention on Biological Diversity (UNCBD,1992) protects biodiversity as a common concern of humankind.[17] Similar to common heritage goods, common concern goods are particularly important to mankind; the concept of common concern of humankind may be even stronger than the concept of common heritage of mankind (Biermann, 1996:476) which was originally developed in the law of the sea. The latter concept had an economic background. Developing countries wanted to ensure that industrialized countries did not exhaust the treasures of the deep sea before the developing countries reached the technical standards necessary to effectively explore the sea. The concept of common concern, to the contrary, can be related to higher natural law principles. Therefore, rules developed for common heritage goods, such as equitable use of a shared resource, should apply even more to common concern goods. In accordance with this assumption, the Convention on Biological Diversity sets forth in its Article 19 (2):

> *Each Contracting Party shall take all practical measures to promote and advance priority access on a fair and equitable basis by Contracting Parties, especially developing countries, to the results and benefits arising from biotechnologies based upon genetic resources provided by those Contracting Parties*
>
> UNCBD, 1992[18]

The United Nations Framework Convention on Climate Change (UNFCCC, 1992), which will be dealt with more extensively in the following section, explicitly states a responsibility of the present generation towards future generations.[19] Otherwise it comprises the usual duties to cooperate, to inform and to assist each other emphasizing the use of shared resources and common concern goods such as the Earth's climate.[20,21] Moreover, the UNFCCC seems to include elements of distributive justice. Article 3 (1) reads:

---

17 See UNCBD (1992) Preamble.
18 See also: Articles 1 and 15.
19 See Article 3 (1).
20 See Articles 3 and 12.
21 According to the preamble of the UNFCCC, the parties acknowledge 'that change in the Earth's climate and its adverse effects are a common concern of humankind'.

> *The Parties should protect the climate system for the benefit of present and future generations of humankind, on the basis of equity and in accordance with their common but differentiated responsibilities and respective capabilities. Accordingly, the developed country Parties should take the lead in combating climate change and the adverse effects thereof.*

The concept of intergenerational equity

The concept of intergenerational equity (Brown Weiss, 1989; 1996) is a modern aspect of equity and of international law in general.[22] Traditional international law deals with relations between states; modern human rights law includes individuals as subjects of international law. However, intergenerational equity goes even further, giving rights to people who are not born yet.

The concept of intergenerational equity has evolved in international environmental law. If some resources are destroyed forever, this may violate our obligations towards future generations. Intergenerational equity is particularly relevant for climate change. Edith Brown Weiss (1989:21) writes:

> *The starting proposition is that each generation is both a custodian and a user of our common natural and cultural patrimony. As custodians of this planet, we have certain moral obligations to future generations which we can transform into legally enforceable norms.*

Accordingly, Article 3 (1) of the Framework Convention on Climate Change commands that '[t]he Parties should protect the climate system for the benefit of present and future generations' (UNFCCC, 1992). Even though formulated in a weak form, intergenerational equity has found its way into a legally binding international convention. We are, thus, witnessing the transformation of 'certain moral obligations to future generations' (Brown Weiss, 1989:21) into legally enforceable norms.

Intergenerational equity and the International Court of Justice

The concept of intergenerational equity has recently found its way into the ICJ. In his dissenting opinion to the order of the ICJ (1995) in the Nuclear Tests Case (*New Zealand versus France*), Judge Weeramantry rightly calls the principle of intergenerational equity 'an important and rapidly developing principle of contemporary international law'.[23] Judge Weeramantry explains that:

> *...this Court must regard itself as a trustee of those rights in the sense that a domestic court is a trustee of the interests of an infant*

---

22 Interestingly, the preamble of the International Convention for the Regulation of Whaling refers to this concept as early as 1946. This convention is reproduced in Birnie and Boyle (1995:586).
23 ICJ (1995:341), Dissenting Opinion of Judge Weeramantry.

> *unable to speak for itself. If this Court is charged with administering international law, and if this principle is building itself into the corpus of international law, or has already done so, this principle is one which must inevitably be a concern of this Court. The consideration involved is too serious to be dismissed as lacking in importance merely because there is no precedent on which it rests.*
>
> *New Zealand's complaint that its rights are affected does not relate only to the rights of people presently in existence. The rights of the people of New Zealand include the rights of unborn posterity. Those are rights to which a nation is entitled and indeed obliged to protect.*

He also refers to the Stockholm Declaration on the Human Environment which formulated, nearly a quarter of a century ago, the principle of 'a solemn responsibility to protect and improve the environment for present and future generations' (UN, 1972: Principle 1). He stated that:

> *...[t]his guideline sufficiently spells out the approach to this new principle which is relevant to the problem the Court faces of assessing the likely damage to the people of New Zealand. This Court has not thus far had occasion to make any pronouncement on this developing field. The present case presents it with a preeminent opportunity to do so, as it raises in pointed form the possibility of damage to generations yet unborn.*
>
> ICJ, Dissenting Opinion 1995:342

This is taken from a dissenting opinion to a judgement of the ICJ. However, dissenting court opinions are often the means through which new legal concepts gain ground and possibly develop into mainstream concepts.

The Philippines: Minors OPOSA versus Secretary of the Department of Environment and Natural Resources (DENR) – a Domestic Court Decision as Evidence of Customary International Law in statu nascendi[24]

According to Article 38 (1) Statute ICJ, customary international law is comprised of two elements: the objective 'general practice' and the subjective 'accepted as law', the so-called *opinio iuris* which may be inferred from the behaviour of states. In order to find out about the precise quality of intergenerational equity, therefore, one has to look at state practice. State practice does not only mean the behaviour of states in their relations with each other.[25] Rather, it also includes a state's legislation and its judicial decisions. Both international and domestic courts play an important role in the application of custom, the decisive element of customary international law (Malanczuk, 1997:39; Pereira and De Quadros, 1995:160; Delbrück and Wolfrum, 1989:58).

In this context it is interesting that the Supreme Court of The Philippines extensively dealt with and recognized the principle of intergenerational equity.

---

24 See DENR (1994).
25 But see Kimminich (1997:221).

Its case OPOSA versus DENR deals with the complaint of a large family against increasing deforestation of The Philippines. Invoking the concept of intergenerational equity, plaintiffs alleged that deforestation violates their yet unborn posterity's right to environmental protection. The trial court dismissed the complaint, ruling that complainants had failed to state a cause of action. The Supreme Court reversed this decision, ruling that 'the plaintiffs have standing to represent their yet unborn posterity, that they had adequately asserted a right to a balanced and healthful ecology' (DENR, 1994:173). The Supreme Court considered a 'right of Filipinos to a balanced and healthful ecology which the petitioners dramatically associate with the twin concepts of intergenerational responsibility and intergenerational justice' (DENR, 1994:175) and finally based petitioners':

> ...*personality to sue on behalf of the succeeding generations ... on the concept of intergenerational responsibility insofar as the right to a balanced and healthy ecology is concerned ... Needless to say, every generation has a responsibility to the next to preserve that rhythm and harmony [of nature] for the full enjoyment of a balanced and healthful ecology.*
>
> DENR, 1994:185

This decision reflects the growing acceptance of the concept of intergenerational equity. 'OPOSA has given a substantial boost to the growing legal legitimacy of the environmental rights of future generations' (Allen, 1994; Rest, 1994) and has shown that the interests of future generations can be identified and advocated by a legal representative.

As domestic courts also contribute to the creation of customary international law (Bernhardt, 1992), OPOSA versus DENR is another argument for the thesis that intergenerational equity is customary international law *in statu nascendi*.

Content and evaluation of equity in international law

Equity under modern international environmental law has several aspects. First of all, there is the principle of equitable burden-sharing between essentially equal partners. According to this principle, the burdens and costs of environmental protection are to be shared between the states on the basis of rational criteria rather than arbitrarily or at random. Moreover, there is the principle of equitable use which applies to shared resources, common heritage and common concern goods. It derives from international customary law as declared and concretized in numerous provisions of international treaties concerning the protection of the environment. This principle implies a duty to assist those states who are unable to participate and share in the resource for financial or technical reasons.[26] In this context, the principle of particularly taking into account the needs of developing countries is increasingly accepted. The principle of equitable use, thus, has some distributive or corrective element.

---

26 Cf UNCBD (1992: 830) Article 20 (2), (3), (4), (5); UNFCCC (1992: 849, 854, 858) Articles 3 (1), (2), 4 (3), (4), (7), (9), 5 (c).

A further step in developing this type of equity in international environmental law is recognizing intergenerational equity as customary international law. Intergenerational equity has only recently found its way into a legally binding international document (Article 3 (1), FCCC). However, in exceptional cases, customary international law may develop quickly. This especially applies when new problems arise which need to be resolved immediately. Then, states might, under pressure, reach consensus. Climate change is such an extraordinary case. Depletion of the ozone layer and climate change cannot be remedied once they occur, at least not easily. This might be one of the reasons why so many states acceded to the FCCC so rapidly. At present, it is probably correct to consider intergenerational equity as customary international law *in statu nascendi*.[27]

Finally, there is equity in the sense of fairness. Such equity is connected with the duty of states to inform each other, to cooperate and to assist each other.[28,29] These procedural duties are to ensure that those potentially affected by a decision are fully apprised of its implications and are given an opportunity to respond. A broad notion of equity or fairness so far implies a form of procedural due process which dictates that those individuals or states likely to be affected by proposed activities should be given a chance to participate in the decision-making process and have their interests or concerns taken into account (Okowa, 1996:278,285).

LESSONS FOR CLIMATE CHANGE

The Vienna Convention on the Protection of the Ozone Layer, the Montreal Protocol on Substances that Deplete the Ozone Layer, and the Framework Convention on Climate Change all contain the usual duties applying to shared resources, common heritage and common concern goods: duties to inform each other, to cooperate and to assist each other.[30,31,32] Moreover, they contain numerous references to the special needs of developing country parties. All these rules are emanations of the principle of equity, as shown above.

The Intergovernmental Panel on Climate Change (IPCC), an expert committee established in the context of the FCCC, dealt with equity in its

---

27 See Rio Declaration (1992) Principle 6. See also UNCLOS (1982) Articles 140 and 203.
28 Antarctic Treaty (1959), Article 3 (1) (a); Antarctic Convention Article 16 (Availability ... of Data and Information); UNCBD (1992) Article 17 (Exchange of Information); Rio Declaration (1992), Principles on Notification: Articles 18, 19; UN Watercourses (1992) Articles 6, 13 (Exchange of Information).
29 Antarctic Treaty Article 3 (1) (International Cooperation); Montreal Protocol (1987); Montreal Protocol (1990) Article 4 (Cooperation): 1 + 2; UN Watercourses (1992) Articles 2, 5, 9, 12 (Cooperation); UN Watercourses (1997) Article 8; NIEO (1974) Article 4. (b); UN Charter (1974) Preamble, paras. 2, 4, 5 (c), 8; Introduction lit. (n) + Articles 3, 13 (4), 17, 23; UNCBD (1992) Article 18 (Technical and Scientific Cooperation); Rio Declaration (1992) Principles on Cooperation: 5 (poverty), 7 (Earth's Ecosystem), 9 (Exchange of Scientific and Technological Knowledge), 12 (Open International Economic System), 14 (Prevent Environmental Harm); GEF (1994) Basic Provisions I 2 + Annex D.
30 Vienna Convention (1985) Articles 3 (3), 5; Montreal Protocol (1987) Articles 7 and 9; UNFCCC (1992) Articles 4 (1), (2), 12.
31 Vienna Convention (1985) Articles 2 (2), 4; UNFCCC (1992) Article 4 (1).
32 For duties to assist and the needs of the developing countries, see Montreal Protocol (1987) Articles 5, 10 (1); UNFCCC (1992) Articles 3 (1), (2), 4 (2a), (3), (4), (7), (9), (10), 5 (c).

second report on economic and social dimensions of climate change, referring among other issues to:

- procedural equity based on the participation and equal treatment of all;
- equitable results: compensation of particular vulnerabilities to climate change and of differentiated capacities to reduce climate change;
- intergenerational equity (and equity within countries); and
- distribution of abatement costs and establishment of emission quotas per capita (IPCC, 1996; Banuri et al, 1996:79,85,118).

The Kyoto Protocol (1997) imposes quantified emission limitation and reduction commitments on developed country parties and parties 'undergoing the process of transition to a market economy' for the period from 2008 to 2012.[33] The primary responsibility of developed parties and parties undergoing the process of transition to a market economy under the FCCC and the Kyoto Protocol complies with the principle of sustainable development, as well as with principles of fairness and equity. Developing countries, and this generally means countries with relatively low per capita emissions, are entitled to a certain 'catching up'. This is true even though, in the long run, developing countries need to stabilize or reduce their emission rates too.

In view of the need to establish further commitments for subsequent periods and with respect to the involvement of developing countries, it remains interesting to find out whether equity might have a more specific and concrete meaning under climate change law. Such considerations might help to establish guidelines for determining greenhouse gas reduction quotas for individual states or groups of states in the future.

Equity in the sense of the United Nations Framework Convention on Climate Change as a basis for reduction quotas

The ultimate objective of the Framework Convention on Climate Change is to achieve 'stabilization of greenhouse gas concentrations in the atmosphere at a level that would prevent dangerous anthropogenic interference with the climate system'.[34] However, the FCCC imposes no precise obligations on state parties. Article 4 (2a) and (2b) only lays down vague obligations for developed countries to reduce their levels of greenhouse gas emissions.[35] This leaves room for equity as a rule of interpretation or gap filler.

The starting point for determining reduction quotas is the often quoted Article 3 (1) of the FCCC. According to that provision, '[t]he parties should protect the climate system for the benefit of present and future generations of humankind, on the basis of equity and in accordance with their common but differentiated responsibilities and respective capabilities. Accordingly, the developed country parties should take the lead in combating climate change and the adverse effect thereof' (UNFCCC, 1992). Article 3 (2) stresses the specific needs and special circumstances of developing country parties.

---

33 Article 3 (7).
34 UNFCCC (1992) Article 2.
35 On specific reduction quotas in preceding treaties, see Sulphur Protocol (1994); CEC (1988).

The convention, thus, integrates the two modern approaches of intergenerational equity and of differentiated responsibilities. Differentiated responsibilities correspond to differentiated contributions to atmospheric pollution in the past and, thus, harmonize with the principle of causation or with the polluter pays principle. On this basis, the – non-binding – Berlin Mandate (1995) of the parties assumes special responsibilities of industrialized states, whereas it excludes additional obligations for the other parties. Those other parties are granted the right to sustainable development which may include certain increases in emissions. According to the Berlin Mandate, state parties and the European Union will elaborate concrete emission limits for industrialized states.

Several criteria can be considered in order to define obligations to reduce greenhouse gas emissions more precisely. Those criteria, however, must be compatible with the principle of equity.

Acquired rights as an equitable criterion?

Reduction obligations could be based on the emission standards reached by states at the time when the Framework Convention on Climate Change was adopted. But such an approach would reward high emission standards in the past. As a result, such a criterion of 'acquired rights' would contradict the objective of the convention to reduce emission standards, and in particular to reduce the high emission standards of industrialized countries. Therefore, this approach does not comply with the requirement of equity in the sense of the FCCC.

Polluter pays principle and principles governing transboundary pollution

An opposite approach would stress the principle of causation or of polluter pays, taking fully into account past emissions. It would be difficult to retroactively impose obligations deriving from the new concept of climate change as a common concern of states. However, the prohibition of substantial transboundary pollution, which is connected with the principle of good neighbourliness, has been generally recognized under traditional international law.[36] The object and purpose of environment protection norms support a broad understanding of the term neighbour. Neighbouring states in this sense are not only the immediate neighbours, but all states affected by the pollution. As climate changes resulting from high emissions of certain gases harm all states, every state is a neighbour in this context (Vitzthum, 1997). This harmonizes with the concept of climate as a common concern and with climate protection as an *erga omnes* obligation under public international law.

Greenhouse gas emissions, however, were not considered as transboundary pollution contrary to international law in the past. Consequently, states would not agree to undergo disadvantages based on those past emissions. It would be unfair and unequitable to take past emissions as a main criterion for the determination of greenhouse-gas reduction obligations.

---

36 Since the Trail Smelter Arbitration(s), 33 *AJIL* 182 (1939) and 35 *AJIL* 684 (1941).

Per capita approach

Modern international law is characterized by the growing importance of *ius cogens*, *erga omnes* obligations and by the emerging role of the individual as subject of international law. It harmonizes with these new developments in international law to take the individual (the number of inhabitants of a state) as the main criterion for determining concrete reduction obligations. This particularly applies in the absence of other equitable criteria. The standard of living and development chances of the individual cannot be isolated from the emission standards in the country where he or she lives. Development presupposes certain emission levels.

The per capita approach further corresponds to the 'differentiated responsibilities and respective capabilities' of states.[37] It burdens industrialized states, with their currently higher emission levels, more than developing states. Moreover, it should be assumed that reductions can be implemented more easily in states with high per capita emissions than in states with low per capita emissions.

Finally, the issue remains whether to take the present population of states as a criterion for their reduction obligations or whether to choose a dynamic approach, taking into account future population growth (and potential decrease). Equity gives no answer to this question. Choosing the present numbers of inhabitants of states as a criterion might have the desirable side effects of a reasonable population policy, including the education of women.

Starting from the per capita approach, the principle of equity takes into account all relevant factors of the specific case. This may lead to modifications of the per capita approach according to factors such as the cooler or warmer climate prevailing in a particular state or its reliance or dependence upon highly emitting industry. Drastic changes in the number of a state's inhabitants may also require long-term adaptations of emission rights, even on the basis of an otherwise static per capita approach.

Moreover, the full realization even of a modified per capita approach cannot take place suddenly. However, its gradual realization 'should be achieved within a time frame sufficient to allow ecosystems to adapt naturally to climate change'.[38]

CONCLUSIONS

Equity is an established principle in modern international law, particularly in environmental law; even intergenerational equity is international law *in statu nascendi*. It implies duties of states to inform each other, to cooperate and to assist each other in the use of shared resources, common heritage and common concern goods. A more recent tendency is the accentuation of the specific needs and special circumstances of developing countries. The 1992 Framework Convention on Climate Change includes such modern concepts within an internationally binding document – namely, intergenerational equity and the principle of 'differentiated responsibilities and respective capabilities' of states.

---

37 UNFCC (1992) Article 3 (1).
38 UNFCC (1992) Article 2.

Under the FCCC, states have differentiated responsibilities to prevent dangerous anthropogenic interference with the climate system. In that respect, this chapter comes to the conclusion that a per capita approach remains the only reasonable and equitable basic criterion for determining reduction obligations. However, equity allows and requires that, starting from this basic criterion, all circumstances of the particular case or state will be considered. In the long run, emission rights of states should, thus, roughly reflect their numbers of inhabitants. But this result should be modified, taking into account the particularities of each state or group of states.

REFERENCES

Akehurst, M, 1976, Equity and general principles of law, *International and Comparative Law Quarterly* **25**:801

Allen, T, 1994, The Philippine children's case: Recognizing legal standing for future generations, *Georgetown International Environmental Law Review* **6**:713

Antarctic Treaty, 1959, Washington, 1 December 1959, 402 UNTS 71

Banuri, T, Göran-Mäler, K, Grubb, M, Jacobson, H K and Yamin, F, 1996, Equity and social considerations, in: IPCC (1996)

Bedjaoui, M, 1979, *Towards a New International Economic Order*, Holmes and Meier, New York

Berlin Mandate, 1995, Berlin Mandate, UNFCC Conference of the Parties: Decisions Adopted by the 1st Session (Berlin, 7 April, 1995), *International Legal Materials* **34**:1671

Bernhardt, R, 1992, Customary international law, in: R Bernhardt, ed, *Encyclopedia of Public International Law* **1**:898

Biermann, F, 1996, Common concern of humankind: The emergence of a new concept of international environmental law, *Archiv des Völkerrechts* **34**(4):426–481

Birnie, W and Boyle, A, 1992, *International Law and the Environment*, Clarendon Press, Oxford

Birnie, W and Boyle, A, eds, 1995, *Basic Documents on International Law and the Environment*, Clarendon Press, Oxford

Brown Weiss, E, 1989, *In Fairness to Future Generations: International Law, Common Patrimony and Intergenerational Equity*, Transnational Publishers, Dobbs Ferry, NY

Brown Weiss, E, 1996, Intergenerational equity and rights of future generations, 601, in: A Cancado Trindade, ed, *The Modern World of Human Rights*, Instituto Interamericano de Derechos Humanos, San José, Costa Rica

Brunnée, J, 1989, 'Common interest' – Echoes from an empty shell?, *Zeitschrift für ausländisches öffentliches Recht und Völkerrecht* [Heidelberg Journal of International Law] **49**:791

Brunnée, J and Toope, S, 1997, Environmental security and freshwater resources: Ecosystem regime building, *American Journal of International Law* **91**:26

CEC, 1988, Directive of the Council of the European Community on the Limitation of Polluting Emissions of Large Combustion Plants (14 November, 1988) (88/609/EEC)

Delbrück, J and Wolfrum, R, 1989, *Völkerrecht*, Walter de Gruyter, Berlin

Dellapenna, J, 1994, Treaties as instruments for managing internationally shared water resources: Restricted sovereignty versus community of property, *Case Western Reserve Journal of International Law* **26**:27

DENR, 1994, The Philippines Supreme Court Decision in Minors Oposa versus Secretary of the Department of Environment and Natural Resources (DENR), *International Legal Materials* **33**:173 (30 July, 1993), Syllabus provided by ILM

Franck, T, 1995, *Fairness in International Law and Institutions*, Clarendon Press, Oxford

Frowein, J, 1984, Jus cogens, in: R Bernhardt, ed, *Encyclopedia of Public International Law* 7:327

GATT, 1947, General Agreement on Tariffs and Trade, 30 October, 1947, TIAS No 1700, 55 UNTS 187

GATT, 1979, Differential and More Favourable Treatment, Reciprocity and Fuller Participation of Developing Countries, GATT Doc L/4903 (28 November, 1979)

GATT, 1994, GATT: Multilateral Trade Negotiations Final Act Embodying the Results of the Uruguay Round of Trade Negotiations (Marrakesh, 15 April, 1994), *International Legal Materials* 33:1125

GEF, 1994, Instrument for the Establishment of the Restructured Global Environment Facility as established by the executive board of the UN Development Programme and of the UN Population Fund (Geneva, 16 March, 1994), *International Legal Materials* 33:1273

Hendel, J, 1996, Equity in the American courts and in the world court: Does the end justify the means?, *Indiana International and Comparative Law Review* 6:637

ICJ (International Court of Justice), 1969, North Sea Continental Shelf Case (*FRG versus Denmark/Netherlands*), *ICJ Report* 3

ICJ (International Court of Justice), 1970, Barcelona Traction Case (*Belgium versus Spain*), *ICJ Report* 3

ICJ (International Court of Justice), 1982, Continental Shelf Case (*Tunisia versus Libyan Arab Jamahiriya*), *ICJ Report* 18

ICJ (International Court of Justice), 1984, Gulf of Maine Case (*Canada versus United States of America*), *ICJ Report* 246

ICJ (International Court of Justice), 1985, Continental Shelf Case (*Libya versus Malta*), *ICJ Report* 13

ICJ (International Court of Justice), 1986, Frontier Dispute (*Burkina Faso versus Mali*), *ICJ Report* 554

ICJ (International Court of Justice), 1993, Case Concerning Maritime Delimitation in the Area Between Greenland and Jan Mayen (*Denmark versus Norway*), *ICJ Report* 177 (14 June)

ICJ (International Court of Justice), 1995, Request for an Examination of the Situation in Accordance with Paragraph 63 of the Court's Judgement of 20 Dec 1974 in the Nuclear Tests Case (*New Zealand versus France*), *ICJ Report* 288 (Order of Sept 22)

ILA (International Law Association), 1966, Helsinki Rules on the Uses of Waters of International Rivers, *ILA Conference Report* 52:484

ILA (International Law Association), 1986, Seoul Declaration on the New International Economic Order, *ILA Report* 5

ILC (International Law Commission), 1996, *Draft Articles on State Responsibility*, GA A/CN 4/L p 28/Add 2, 16 July, 1996, GA OR, 51 Suppl 10, 125

IPCC (Intergovernmental Panel on Climate Change), 1996, *Climate Change 1995, Economic and Social Dimensions of Climate Change*, Contribution of Working Group III to the Second Assessment Report of the IPCC, Cambridge University Press, Cambridge, UK

Janis, M W, 1995, Equity in international law, in: R Bernhardt, ed, *Encyclopedia of Public International Law* 2:109

Kimminich, O, 1997, *Einführung in das Völkerrecht*, A Francke Verlag, Tübingen

Kiss, A and Shelton, D, 1991, *International Environmental Law*, Transnational Publishers, Ardsley-on-Hudson

Kokott, J, 1998a, Grund- und Menschenrechte als Grundlage eines internationalen ordre public, *Berichte der Deutschen Gesellschaft für Völkerrecht* 38:71

Kokott, J, 1998b, Überlegungen zum völkerrechtlichen Schutz des Süßwassers, p177 in: R Wolfrum, ed, *Liber arnicorum Günther Jaenicke – zum 85 Geburtstag*

Kyoto Protocol, 1997, Kyoto Protocol to the United Nations Framework Convention on Climate Change, *International Legal Materials* 27:22 (1998)

Lachs, M, 1993, Equity in arbitration and in judicial settlement of disputes, *Leiden Journal of International Law*, 6:323

Langer, St, 1995, *Grundlagen einer internationalen Wirtschaftsverfassung*, (Principles of a New International Economic Order), C H Beck, Munich

Malanczuk, P, 1997, *Akehurst's Modern Introduction to International Law*, Routledge, London

McCaffrey, St and Sinjela, M, 1998, The 1997 United Nations Convention on International Watercourses, *American Journal of International Law* **92**:97

Montreal Protocol, 1987, Protocol on Substances that Deplete the Ozone Layer (done at Montreal, 16 September, 1987; in force since 1 January, 1989; ratified by 157 states), *International Legal Materials* **26**:1541

Montreal Protocol, 1990, Adjustments and Amendment to the Montreal Protocol (London, 29 June, 1990; Amendment in force since 10 August, 1992; ratified by 109 states), *International Legal Materials* **30**:537

NIEO, 1974, Declaration on the Establishment of a New International Economic Order (adopted by GA on 1 May, 1974 as Res 3201 (S-VI) without vote on its 6th Special Session), *International Legal Materials* **13**:715

Okowa, P, 1996, Procedural obligations in international environmental agreements, *British Yearbook of International Law* **67**:275

Oxman, B, 1997, The international commons, the international public interest and new modes of international lawmaking, p 21 in J Delbrück, ed, *New Trends in International Lawmaking – International 'Legislation' in the Public Interest*, Proceedings of an International Law, 6–8 March, 1996

Pereira, A G and De Quadros, F, 1995, *Manual de Direito Internacional Publico*, Livraria Almedina, Coimbra

Rest, A, 1994, The OPOSA Decision: Implementing the principles of intergenerational equity and responsibility, *Environmental Policy and Law* **24**:314

Riedel, E, 1997, International environmental law – A law to serve the public interest? – An analysis of the scope of the binding effect of basic principles (public interest norms), p 61 in J Delbrück, ed, *New Trends in International Lawmaking – International 'Legislation' in the Public Interest*, Proceedings of an International Law Symposium of the Kiel Walter-Schlücking Institute of International Law, 6–8 March, 1996

Rhinesalt Convention, 1976, Convention on the Protection of the Rhine River against Chemical Pollution and on the Protection against Pollution through Chlorides (Rhinesalt Convention), 3 December, 1976

Rhinesalt Convention, 1991, Protocol concerning the Rhinesalt Convention 25 September, 1991

Rio Declaration, 1992, Rio Declaration on Environment and Development, *International Legal Materials* **31**:876

Rossi, C, 1991, *Equity as a Source of Law? A Legal Realist Approach to the Process of International Decision-Making*, UMI, Ann Arbor, MI

Sulphur Protocol, 1994, Protocol on Further Reduction of Sulphur Emissions (Oslo, 14 June, 1994) *International Legal Materials* **33**:1542

UN, 1972, Declaration of the United Nations Conference on the Human Environment (Stockholm, 16 June, 1972), *International Legal Materials* **11**:1416

UNCBD, 1992, United Nations Convention on Biological Diversity (Rio de Janeiro, 5 June, 1992; in force since 29 December, 1993; ratified by 175 states as of 15 January 1999), *International Legal Materials* **31**:818 (1992)

UN Charter, 1974, Charter on Economic Rights and Duties of States (adopted by GA on 12 December, 1974 as Res 3281 (XXIX), by a vote of 120 in favour to 6 against, with 10 abstentions), UN Doc A/Res/3281 (XXIX), *International Legal Materials* **14**:251 (1975)

UNCLOS, 1982, United Nations Convention on the Law of the Sea, *International Legal Materials* **21**:1261

UNFCCC, 1992, United Nations Framework Convention on Climate Change (New York, 9 May, 1992; in force since 21 March, 1994; ratified by 176 states as of 7 October 1998 (see also Multilateral Treaties Deposited with the UN Secretary-General, http://www un org/Depts/Treaty), *International Legal Materials* **31**:849

UN Watercourses, 1992, United Nations Convention on the Protection and Use of Transboundary Watercourses and International Lakes (Helsinki, 17 March, 1992, in force since 23 October, 1996, ratified by 19 states), *International Legal Materials* **31**:1312

UN Watercourses, 1997, United Nations Convention on the Law of the Non-Navigational Uses of International Watercourses, Adopted by GA and opened for signature on 21 May, 1997, *International Legal Materials* **36**:700

UNEP (United Nations Environment Programme), 1978, Governing Council['s] Approval [by consensus] of the Report of the Intergovernmental Working Group of Experts on Natural Resources Shared by Two or More States on 19 May, 1978, Doc UNEP GC 6/CRP 2, *International Legal Materials* **17**:1091

Vienna Convention, 1969, Vienna Convention on the Law of Treaties (adopted on 22 May, 1969, entered into force on 27 January, 1980), UN Doc A/CONF 39/27, 289

Vienna Convention, 1985, Vienna Convention for the Protection of the Ozone Layer (Vienna, 22 March, 1985; in force since 22 September, 1988; ratified by 160 states), *International Legal Materials* **26**:1516

Vitzthum, W, 1997, *Völkerrecht*, p 485, in: W Graf Vitzthum, ed, *Völkerrecht*, Walter de Gruyter, Berlin

WTO, 1994, Agreement Establishing the World Trade Organization (in force since 1 January, 1995; acceded by 130 states), *International Legal Materials* **33**:1125

12 THE REGULATION OF GREENHOUSE GASES: DOES FAIRNESS MATTER?

David G Victor[1]

INTRODUCTION AND OVERVIEW

The search for fair or equitable agreements to slow global warming has become a cottage industry. Dozens of papers have proposed solutions. International negotiations to strengthen the UN's Framework Convention on Climate Change (FCCC) regularly consider the topic. The Kyoto Protocol's commitments for industrialized countries to regulate emissions of greenhouse gases are 'differentiated' to each country's particular situation, which appears to be evidence that negotiators have taken seriously the need for fairness. Of special interest to the organizers of this workshop is whether fairness should be a major element of research programmes that employ integrated assessment models – models that allow an integrated analysis of the costs and benefits of policy options for slowing global warming.[2]

Conventional wisdom holds that fair agreements are more effective; if so, integrated assessment models that are oriented towards assessing the fairness of policy options would be important tools for negotiators and other policy-makers. Thus, fairness is not only a crucial issue for the design of international agreements but also for the design of policy-relevant research programmes.

This chapter argues that the conventional wisdom is largely incorrect. Firstly, it suggests that fairness has little influence on the substantive commitments in international environmental agreements, especially when

---

1 The author has benefited enormously from work with colleagues at IIASA in writing and editing a book that includes 14 case studies on implementing international environmental commitments (see Victor et al, 1998). Thanks are also due to reviewers, Dan Bodansky, Ferenc Tóth and Farhana Yamin, for comments on an earlier draft. This is intended as a provocative chapter on a topic that is sensitive but crucial for policy-relevant research; the views expressed here are those of the author only.

2 The hosts, Potsdam Institute for Climate Impacts Research (PIK), convened the workshop to discuss the fairness question. PIK is in the midst of developing an integrated assessment model through a partnership with many other research institutions. One contemplated use of the model is to assess the fairness of potential international agreements to regulate emissions of greenhouse gases.

commitments require governments to implement costly actions.³ Most of the evidence that has been cited to support the proposition that fairness is important actually illustrates a different (but correlated) factor at work: variations in willingness to pay for international environmental regulation.⁴

Secondly, this chapter notes that fairness is but one aspect of the problem of how to differentiate commitments among the participants in international agreements. In practice, it is difficult to negotiate differentiated commitments, and thus even if fairness were important it would be difficult to put into practice. The primary lesson to be drawn from those few cases where sophisticated models have been used to formulate differentiated commitments is that good models depend critically on widely accepted methods and databases. In the case of climate change, such methods and data do not (yet) exist. Serious differentiation of commitments requires giving priority to building confidence in the analytical tools that will be used to assess costs and benefits – some efforts are underway, but much more is needed. In the negotiations for the Kyoto Protocol, differentiation was pursued only in the final months, with critical aspects of the protocol's commitments settled only in the final hours. A more systematic analysis, rooted in more elaborate analytical methods, will be needed in the future if countries are to implement schemes that require sophisticated differentiation of the commitments.

Taken together, these two conclusions suggest that analysts who make fairness the centrepiece of efforts to identify the elements of a successful international agreement to regulate greenhouse gases will not have much influence on real outcomes. The purpose of this chapter is to elaborate upon and illustrate – mainly with examples from research projects in which the author has participated – an alternative view that has major implications for theory and research. It is not a full-blown scholarly treatment of the issues or literature (for an introduction to the literature, see Banuri et al, 1995).⁵ This chapter does not argue that research on fairness is irrelevant; nor should the cottage industry be closed. However, such analysis must be coupled with sober assess-

---

3 This chapter does not address so-called 'procedural fairness' concepts, such as the need for a policy-making process that is highly participatory, access to information, minority rights, etc.
4 This chapter uses the common-sense definition of the concept of willingness to pay: willingness to devote resources so that behaviour is different from what it would have been otherwise. Willingness to pay corresponds most closely with the present objectives of international agreements, which are to allocate the burden of controlling behaviours that cause greenhouse gas emissions. Willingness to pay is related to willingness to accept climate changes; moreover, if international agreements give greater attention to adaptation to climate change, then willingness to accept may play a larger role. However, in discussions of allocating burden, willingness to pay matters most. For more, see the brief review in Chapter 4 of this book.
5 For a statement of this view, a review of the literature, and the main arguments about the importance of fairness (against which this author will argue in this chapter) see the relevant IPCC chapter (Banuri et al, 1995). This chapter is not intended necessarily as a critique of the IPCC authors – their task was to review the literature. Rather, it is a critique of the assumptions and concepts in the literature. Moreover, this chapter will not consider the relevance of fairness as expressed in decisions by the International Court of Justice (ICJ) and through liability schemes, though the IPCC chapter does address those issues. In this author's view, neither the ICJ nor liability schemes are relevant to adopting and implementing international agreements to slow global warming. Liability schemes are rarely used and require proving harm, which cannot be done for global warming impacts in the foreseeable future (see the final section of this chapter). The ICJ resolves disputes, which are mainly bilateral in nature. ICJ decisions have little if any sway on multilateral environmental negotiations; and the dispute resolution procedures of multilateral environmental agreements have never been invoked and are unlikely to be in the future.

ments of different societies' willingness to pay for greenhouse gas abatement. Where the gap between willingness to pay and fairness is large, the prospects for a fair agreement will be low. Improved capacity to analyse willingness to pay is crucial since it, not fairness, will be the driving force of effective international cooperation. Efforts are underway to improve the utility of integrated assessment models for this purpose. This chapter also suggests two factors that are probably important but are today not part of that research programme – public perceptions of the risk of climatic catastrophe and the spread of liberal (democratic) decision-making. These factors affect how societies view the possible costs of climate change and the ability to express those views in the policy-making process.

Does Fairness Matter?

It is crucial that analysts distinguish normative from positive in order to separate pleas for what should happen from analysis of what is likely to happen. The debate has been dominated by the former; here this chapter focuses on the latter. Studies of 'fair' agreements are typically based on the assumption that, all else equal, a fair agreement is more likely to be accepted than an unfair one. But all else is rarely equal, and thus the attribute of fairness will be traded against other factors, such as costs. Normative debate is the backbone of moral decision-making. But arguments about justice without proper full consideration of other factors can easily lead to research products which have little relevance to real policy decisions. The danger is especially acute at the international level.

Within societies, fairness can and does play a role. Societies consist of people who interact on a regular basis, and share values and heritage. They could imagine performing, for example, the Rawls experiment of sitting together behind a veil of ignorance and making decisions according to principles of justice. Often the borders of cohesive societies also define the borders of countries, which is one reason why there are many examples of fairness guiding the domestic policies in many nations. Social institutions such as welfare, state-subsidized health care and progressive income taxes are all partial examples. However, this chapter does not refer to decision-making within countries. That a village in Switzerland, populated by the same families for centuries, has sustained a fair allocation of scarce pastures says little about the role of fairness when the US, Ghana and 167 other countries decide whether to accept an agreement to slow global warming.

It is not easy to draw the line between a society and collections of people that are not a society. Two criteria help to identify both the boundaries of society and the reasons why fairness can play a significant role within societies. One is that societies typically include explicit transfers of many forms, which help target resources to the less fortunate. The other is that within societies it is difficult or impossible to exit from collective policies; often decisions by liberal democratic governments include some element of fairness, without which policies would not be adopted by the polity. Once policy decisions are made, all subjects within the society are bound (by threat of force) to implement them.

International policy decisions, which often take the form of treaties and other agreements, are not the product of a cohesive society. Nor are all states bound to join decisions they do not like. Rather, they reflect the consent of each of the individual states that constitute the nation–state system. In delivering their consent, these units primarily weigh their interests according to the particular issue at hand. Obviously, issues are linked and relations among states are not purely anarchic; traces of 'international society' do exist, especially in Europe. But for most states most of the time, the decision-making process is mainly a selfish one. Fairness is not irrelevant, but it explains a tiny fraction of the variance.

Nonetheless, international agreements on international environmental problems often impose more costly commitments on rich countries, and increasingly they expect rich countries to make financial transfers to assist less wealthy nations to pay the cost of implementation. Is this evidence of fairness influencing outcomes? No. Rather, the rich have a higher *willingness to pay* for most types of international environmental protection. In part, the environment is a luxury good – its value rises with income, leading the rich to care about a wider range of environmental threats. That is especially true for problems that threaten environmental amenities but have little direct and near-term effect on welfare such as, to some extent, global warming, loss of biodiversity and acid rain.[6] In part, many forms of pollution are higher in rich countries and thus so is environmental damage – the rich care more because they have already imposed much harm on themselves.

High willingness to pay partially explains the outcomes of some of the most prominent examples that have been used by advocates of fairness. It explains why the industrialized countries have agreed to more stringent regulations on ozone-depleting substances than exist for developing countries. Moreover, industrialized countries are compensating developing countries for the cost of complying with the Montreal Protocol. The high willingness to pay by industrialized countries is evident in the fact that virtually all contributions to the protocol's voluntary Multilateral Fund have been paid – even the US, UN debtor *par excellence*, is paying its share.[7] If America did not pay, environmental groups, as well as firms that have invested in ozone-benign technologies, would crucify the government.

Willingness to pay also largely explains outcomes in efforts to protect the Baltic Sea. Western nations are paying part of the cost of cleaning up facilities in the former centrally planned countries. In the West, concern for a clean Baltic has been higher than in the East; and the marginal benefit of cutting pollution from already relatively clean sources in the West is much lower than from less costly cleanup of dirty sources in the East. In Russia, where domestic pressure for cleanup is relatively low and thus willingness to pay is small, Western resources are most important. In Poland, where willingness to pay is

---

6 Many local environmental problems, such as undrinkable water and urban air pollution, have direct economic effects, but they are typically not the subject of significant international agreements. However, such problems are often the subject of some international grant and lending programmes, which underscores that fairness is not irrelevant. People and societies help others in many ways that cannot be explained solely by self-interest.
7 Most of the transition countries have failed to pay – in the midst of turmoil at home they understandably have not devoted hard currency to distant stratospheric benefits. Several countries, including Russia, are now net recipients of money to help limit consumption of ozone-depleting substances.

higher, a much larger fraction of the cost of cleanup is being paid by Polish funds. Indeed, indigenous Polish resources are focused on pollution sources that foul local Polish waters; in contrast, foreign resources account for a larger share of the cleanup costs for pollution sources that foul the Baltic Sea commons. Polish willingness to pay for abatement has been higher where Polish interests have been directly harmed by pollution. Indigenous Polish expenditure on pollution control has also risen because compliance with strict environmental rules is a condition for Poland's entry into the European Union. The full story is more complicated, but the broad contours are defined by willingness to pay, which is a function of the pressure on each country's government – internally from its population and externally from its neighbours who have linked pollution control with other objectives.[8]

Willingness to pay also strongly shaped the European Union's (EU's) burden-sharing agreements for controlling carbon emissions. In preparation for Kyoto, the EU's leaders on global warming (notably, The Netherlands, Germany, Sweden and UK) agreed to stringent cuts (though with little idea how to implement them). They were under pressure from within to do something about the climate problem. In less wealthy EU member states, pressure was lower and therefore so was willingness to accept stringent targets. The package of differential targets was cemented by The Netherlands which, as holder of the rotating EU presidency in early 1997, had a special willingness to pay to avoid a diplomatic collapse and thus accepted a particularly stringent target. Countries with the least internal pressure and largest list of other priorities for resources – such as Greece, Spain and Portugal – probably would have avoided any regulation if not for the inviolable requirement that the EU, a nascent society, take a unified approach.[9] Witness Turkey which, like Greece, faces low internal pressure to slow greenhouse emissions but, unlike Greece, is not compelled to follow a common EU approach. Turkey refused any commitments under the Kyoto Protocol and has also asked to have its name expunged from the FCCC, which it never signed. In Russia, where pressure is low and Western entanglements are few, the government is poised to adopt the Kyoto Protocol only because the protocol gives Russia a lenient target, and Russia can make money by selling excess emission rights. In contrast, countries that are entangled with the West – such as the Czech Republic, Hungary, Poland and Slovenia, which are slated to join the EU – have shown greater willingness to adopt more significant constraints on greenhouse gas emissions. As with the example of Baltic pollution, the West linked the issue to benefits of Western integration.

Willingness to pay is obviously correlated with most notions of fairness. The outcomes are similar – rich people pay more – but the causal mechanisms are quite different. The argument that fairness matters depends on the belief that what is perceived to be fair will exert an influence because it is fair – that morality will sway the outcome. When viewed through the lens of willingness to pay,

---

8 For the full story, see Roginko (1998) and Hjorth (1998), and references therein.
9 Nonetheless, that unity will be weaker than initially appears – a special provision in the Kyoto Protocol (Article 4.6) will make each EU member rigorously responsible only for its own emission target, rather than placing responsibility collectively with each EU member state. At this writing, the EU targets are being renegotiated. Before Kyoto the target summed to a 10 per cent to 15 per cent cut in carbon emissions; however, in Kyoto the EU was required only to cut 8 per cent below 1990 levels (of a basket of gases, not just carbon). Probably there will be some relative adjustment in the renegotiation – the Dutch, for example, may be less willing to pay an extra burden because they no longer hold the EU presidency.

the causal mechanism is calculation of self-interest. The differences are important because they help to explain national choices. For example, Germany, which has a strong interest in the EU, pays the most per head to the EU budget; yet several countries (Austria, Belgium, Denmark, France, and Luxembourg) have higher per capita incomes. Belgians are a bit wealthier on average than Germans but pay only a fifth, per head, of the EU budget.[10]

The implications for researchers, such as the developers of integrated assessment tools, is that we should focus more on the elements of self-interest and less on the hope that fairness – whatever that may be – will prevail, divorced of self-interest. Of course, self-interest is increasingly not a purely financial concept, especially when ecological issues are at stake. Many schemes to save biodiversity cost more than their financial returns to society, yet they command strong support. Much of the same can be said for most efforts to limit acid rain and many proposals to slow global warming. People often value the environment beyond its instrumental benefits to humans. The point is that by developing a means to track how societies value environmental protection, one will have a device for predicting willingness to pay and thus a device for predicting what types of agreements can earn consent and be effective.

Although more relevant to reality, this proposal to focus on willingness to pay rather than on fairness is not easy to implement. Several modelling efforts are underway (see, Chapter 4). Here the text highlights two factors are highlighted here that are not included in current modelling efforts but may be crucially important to accurate assessments of the willingness of societies and governments to slow global warming. In turn, such assessments can help to analyse whether likely outcomes will correspond with a 'fair' agreement. If the correspondence is low, then the prospects for fairness are grim.

One factor is the perception of the problem. In the case of global warming, it could be fruitful to evaluate whether societies view the problem as a looming catastrophe or as a threat to normal economic activity. For the former, willingness to pay for abatement is probably high, in part because catastrophe can be costly, and in part because it is probably very difficult to marshall evidence that fully dispels catastrophic fears. For the latter, evidence is easier to weigh and thus action to slow global warming is more readily traded against other social problems and policies. The difference in view partially explains why rich Western countries, which are generally the least vulnerable to global warming, have been the strongest advocates for slowing global warming. They fear a possible (although unlikely) climate catastrophe; their willingness to pay will rise as catastrophes becomes more tangible (as it has in the past, when creative imagination has linked scorching weather, floods and hurricanes to global warming).

A second aspect of evaluating willingness to pay may be the spread of liberal democracy.[11] As more societies open their political systems, their decisions will be affected by a wider array of interests. Those that have been

---

10 Net contributions to the EU budget are not published, which perhaps illustrates that perceptions of fairness do matter within the European society. My assessment here is based on data compiled by the *The Economist* (23 November 1996). Data are adjusted for differences in purchasing power.
11 This chapter will equate democratic with liberal decision-making, though the full story is obviously more complicated. What matters most in this argument is the increased access to information as well as levers on policy decisions to individuals throughout societies, not only to elites. For more on how the spread of liberalism affects the making of international agreements, see Slaughter (1995).

historically excluded, such as environmental groups, will probably enjoy relative gains in influence. The result need not be stricter regulation of greenhouse gas emissions – environmental interests, where they exist in previously undemocratic countries, have many items on their agenda. But the environmental agenda, like many ideas, is increasingly a worldwide phenomenon – and global warming is on that agenda. Moreover, policy action to address some of the environmental issues that are on the agendas of democratizing countries, such as urban smog, will also limit emissions of greenhouse gases. With democratization, international agreements will increasingly reflect the interests of societies; governments will increasingly be mere agents whereas previously they could behave more as unitary actors. The change requires new tactics. Countries that want global action to slow global warming will find it efficient to spend money subsidizing like-minded pressure groups (NGOs) in laggard (but democratic) countries. Doing so will increase the laggards' willingness to pay. Such actions will probably do more to gain the consent of laggards than devoting comparable resources to making an agreement that all consider fair. Since many environmental leaders already subsidize NGOs in laggard countries, it seems that decision-makers understand this point.

THE PROBLEM OF DIFFERENTIATION IN INTERNATIONAL LAW

The second aspect of the fairness debate considered here is how to codify international agreements. Whether one calls it fairness or willingness to pay, the fact remains that the willingness of countries to consent to international commitments will differ because countries are different. Unless commitments merely mirror the interests of the least ambitious country, the need for differentiated commitments is inevitable.

Experience with differentiated commitments in international environmental agreements is minimal. Typically, commitments are not differentiated. In agreements to limit pollution, across-the-board cuts have been the norm. In agreements on wildlife management, typically all parties have been bound by the same commitments. The Montreal Protocol, which is often seen as a vanguard of differentiation, has formally distinguished commitments for only two broad groups of countries – developed and developing. The benefits-sharing scheme for deep seabed resources, negotiated as part of the Law of the Sea, illustrates a more sophisticated formulaic approach; however, that scheme was easier to agree upon because no commercial deep seabed mining existed at the time and thus few entrenched interest groups were inclined towards or against any particular proposal (nonetheless, those negotiations took many years to complete). In contrast, negotiations to slow global warming are being conducted in an atmosphere supersaturated with interest groups who are keen to minimize their cost of complying with an international agreement. Other areas of international cooperation, such as dividing the cost of the UN or the North Atlantic Treaty Organization (NATO), illustrate more sophisticated schemes for differentiation, although both were founded in the era of US hegemony, when the US was willing to pay extra costs which made it easier for all others to agree.

Why do parties to international agreements adopt such simple-minded approaches? Moreover, why do they do so when across-the-board measures appear even handed but, in fact, impose different costs because countries have different starting positions and economies? In part, the answer is simplicity. International negotiations are extremely complex; and deprived of a compelling alternative, the simplest formula is adopted. Moreover, many international commitments do not actually require much additional effort. Differentiation is only important when commitments require some significant action. Climate change now falls into that latter category, as evident in the differentiated targets of the Kyoto Protocol; however, there is relatively little historical experience to guide the effort because few other international environmental agreements have entailed substantial changes in behaviour.

Because most international commitments have been broad and modest, in practice much of the differentiation in commitments has occurred during the process of implementation. In some instances, the exact requirements of international commitments were actually negotiated after the commitments were adopted – the line between negotiation and implementation is increasingly fuzzy. For example, Russia and a few other countries are formally not in compliance with the Montreal Protocol. Through the protocol's implementation committee, they have negotiated a package of domestic measures and international financial assistance (from Western nations through the Global Environment Facility), along with regular performance reviews, which will result in these countries eventually complying with the protocol. Nearly all other industrialized countries have already complied. Many cut consumption of ozone-depleting substances more rapidly than required because they faced strong domestic pressure from environmental groups, and some industrial producers of the offending substances, to implement swift action. Therefore, while all industrialized countries adopted the same commitments, in practice they have actually implemented different commitments.[12]

While much differentiation occurs during implementation, it is possible to negotiate international agreements that include complicated burden-sharing arrangements. The 1994 protocol to the European agreement to cut sulphur emissions in Europe on long-range transboundary air pollution (LRTAP) is an often-mentioned example. In that case, computer models were used to frame negotiations about how much each country should cut its emissions by. The case illustrates that when there is a widely shared metric for assessing complicated goals, differentiation of commitments is possible. In this case, the shared goal was the need to make abatement decisions on the basis of ecological critical loads. (The final numbers were adjusted, by negotiation, to accommodate the lower willingness to pay for costly abatement in the former centrally planned countries, though many of those countries already had low emissions due to economic collapse.) This experience also illustrates that numerical differentiation of commitments requires confidence in numbers. The models used for the 1994 sulphur protocol were built on more than a decade of active efforts to improve data reporting and modelling.[13]

---

12 For additional discussion and illustrations of differentiation of broad commitments through implementation, such as in the North Sea accords, see Raustiala and Victor (1998).
13 For more detail see Levy (1995); and, for additional discussion of the lessons to be drawn from the experience, see Victor and Salt (1995a).

In sum, international environmental agreements have been blunt instruments and thus often poorly suited as devices for imposing differentiated rights and responsibilities on nations. Except in a few cases, they have also been a poor means of facilitating the complicated bargains that will be needed to achieve demanding international cooperation – that is, cooperation that imposes significant burdens and requires accommodating many varied interests. Most differentiation has been through implementation, but relying upon that route as the primary means of differentiation is probably feasible only when the stakes of international cooperation are low. Therefore, it is not necessary for international cooperation to codify all the details of differential obligations into law. This state of affairs does not bode well for the rule of law, nor is it promising for efforts to achieve fairness through fine differentiation of commitments. Some implications follow for research and policy.

Firstly, international negotiations have erred strongly on the side of producing simple agreements. Thus researchers investigating fairness issues should compare schemes that formally differentiate commitments with those that employ across-the-board measures. How much is gained by employing complicated differentiation schemes?

Secondly, if one thinks that complicated numerical burden-sharing arrangements might be valuable in the future, then the single most important thing to do now is to ensure that data reporting and review procedures begin promptly and vigorously. This is the most important lesson to draw from the experience with the 1994 sulphur protocol experience. This point has been argued elsewhere in this chapter and the author has also suggested that such an activity is more important than pursuing targets and timetables for abating greenhouse emissions, especially because the willingness to pay for abatement is (currently) low.[14] So far the reporting and review system is off to a good start, but there are disturbing signs that some countries are modifying data to make their compliance look more favourable. Moreover, data on policies and measures are not comparable across countries and are generally of low quality. Comparable and dense data sets are necessary to allow the mutual scrutiny of policy proposals, which is essential if nations are to negotiate, adopt and implement meaningful targets for global warming that are differentiated.

Thirdly, research and policy on the differentiation of commitments must focus on the process of implementing and not only on negotiating agreements. Implementation is what really determines the degree of differentiation.

THE KYOTO PROTOCOL

The Kyoto Protocol suggests that the international negotiations on global warming may be at a crossroads. The protocol includes targets that are differentiated, which illustrates that differentiation is possible. A surface glance at the protocol also suggests that differentiation was used to promote fairness. On both scores, it is too early to make an assessment. The differentiated targets are evidence that the Kyoto Protocol is the first formal step down the path of implementing significant commitments to limit greenhouse gas emissions. As the collective goal has become more costly, the need for differentiation to

---

14 See two papers: Victor and Salt (1995b) and Victor and Salt (1995a).

account for national differences has risen. Therefore, the Kyoto Protocol is formally differentiated, while the earlier and more modest FCCC included only across-the-board commitments to control emissions (for industrialized countries), with differentiation only through the process of implementation (for example, Sweden did more to control emissions than Austria, although both had adopted the same international commitments).

Below the surface, fairness is much less evident in the Kyoto Protocol. Rather, as argued in this chapter, the differentiation largely reflects willingness to pay. Australia, which faced weak domestic pressure to cut its emissions, feared potentially high costs from regulating its energy-intensive mining industry, and was represented by a government hostile to any agreement that imposed costs on its nation. It was granted an 8 per cent increase. Neighbour New Zealand agreed to freeze its emissions – it accepted a tighter limit because it has plenty of space for carbon-absorbing trees and plans to make money by selling credits for the excess carbon sopped up by its managed forests. Iceland, which faced almost no domestic pressure and did not want Kyoto limits to interfere with its ambitious industrial development plans that include large new aluminium smelting plants, was granted a 10 per cent rise.

Perhaps the most puzzling case is the US, which agreed to a substantial (7 per cent) cut. Special wording in Article 3 of the Kyoto Protocol means that the 7 per cent cut is equivalent to only, perhaps, a 3 per cent cut; nonetheless, the reduction from the unregulated level of US emissions is substantial, perhaps more than 30 per cent. Although US environmentalists pushed for deep cuts and thus raised the willingness of the US to pay for reductions, this substantial commitment might be seen as evidence that the world's largest emitter of greenhouse gases accepted the need to demonstrate fairness by taking the lead. This anomaly is in part evidence that the US delegation went far beyond what the US is willing to pay – few credible observers think that the US can or will ratify the Kyoto Protocol in the near future. Indeed, the process of treaty ratification in the US – controlled by the Senate, that is not necessarily dominated by the same political party as the executive that negotiates treaties – creates a double hurdle of scrutiny of the country's willingness to pay. The US administration is also relying upon Russia to sell excess greenhouse permits; in Kyoto, Russia was given the goal of freezing its emissions, but it is plausible that Russian emissions will be lower. Russian credits will lower the cost of complying with the Kyoto Protocol, making the US Congress and public more willing to ratify the Kyoto accord. It is still uncertain whether it will be politically feasible to ratify a protocol that can be complied with only by purchasing a windfall of permits from Russia. The author doubts it.

Crucially, the deal in Kyoto has led the US government to pursue a worldwide campaign to enlist developing countries to commit to regulating their emissions of greenhouse gases. Observers who believe that fairness is essential to a climate treaty often cite the need for rich industrialized countries to cut emissions while developing countries are left alone. That was the deal struck in the 1992 FCCC and again in 1995 with the Berlin Mandate, which gave developing countries immunity from controls on greenhouse emissions. Yet industrialized countries did little to slow emissions. Now facing the prospect of more costly controls, the US is unwilling to pay unless others – including developing countries – take on significant burdens. This outcome is exactly the opposite of what most observers have claimed would be fair. A general proposition may be that strict fairness is easy to adopt when the stakes are low, but as international commitments become more costly the contours of the agreement follow more closely a country's willingness to pay.

Neither the FCCC nor the Kyoto Protocol impose any formal obligation on developing countries to control emissions. The absence of controls on emissions from developing countries reflects that those countries have been nearly unified in their low willingness to pay the costs of slowing global warming. China, which has totalitarian control on public debate within its borders and many items on its agenda more pressing than global warming, has been the most ardent defender of developing country immunity. Its sheer size and control over its constituency has allowed China to identify and pursue its interests vigorously and effectively. It has been able to credibly demonstrate its low willingness to pay. Only when outsiders (eg the GEF) have paid the incremental costs has China been willing to implement global warming projects.

Only a few developing countries may assume possibly significant obligations, and each case is the result of special factors that increase willingness to pay. Korea and Mexico are under strong pressure to adopt significant regulatory obligations – in both cases, sustaining good membership in the OECD club of industrialized nations gives them an extra incentive to demonstrate some effort, even though both have been formally classified as developing countries in other international environmental agreements. Argentina and Chile have signed declarations of their intent to control emissions, although it is unclear whether these entail any action beyond a symbolic effort, orchestrated by the US administration, to make the US Congress more willing to ratify and implement the Kyoto accord. Argentina has supplied top diplomats to the FCCC and Kyoto Protocol processes and hosted the Conference of the Parties in November 1998, which was the first high level follow-up meeting after Kyoto. Chile is keen to join free trade agreements with the US. Thus, both have a special incentive to aid the US domestic political effort and to keep the Kyoto Protocol from derailing – in both cases this has raised their willingness to participate in symbolic action and perhaps also their willingness to pay for substantive controls on emissions. Statements from other countries are likely in the future as the US continues its 'full court press' in developing countries to declare (at least symbolically) their intention to control emissions. A few, led by Argentina, may adopt binding caps on their emissions because such caps could yield a windfall in emission permits that can be sold overseas.

The most important split in the position of developing countries has not been over adopting costly commitments but rather whether to allow industrialized countries to earn greenhouse reduction credits through investments in the developing world. Many developing countries (led by China and the Group of 77 coalition) have opposed this as an effort by the industrialized world to skirt its obligations and a dangerous first step towards saddling developing countries with costly future obligations. Others, especially in Latin America, have viewed this as a means primarily to earn money in exchange for implementing global warming projects that also often yield many local benefits. Developing countries favourable to the concept have been willing to control emissions if the extra costs are paid by others, such as through the concept of joint implementation (now recast and renamed in the Kyoto Protocol as the clean development mechanism). Already many such projects are underway, funded by industrialized countries.[15] But the costs of regulating greenhouse gases will be much higher than in the case of ozone-depleting substances; so the elegant solution of industrialized countries paying the total agreed incremental costs of abatement is unlikely to hold. Rather, financial transfers will be supplemented by diplomatic and trade pressures to increase the willingness of developing countries to pay for climate control. Many observers who are concerned about fairness will find this reprehensible, but in a world where decisions are made on the basis of self-interest (willingness to pay) rather than on a broader moral compass, this is important to have in mind as analysts devise future scenarios.

FINAL THOUGHTS

To close, this chapter speculates on three topics and introduces some caveats.

Firstly, this argument hinges on how societies assess their values, which this author claims is mainly a selfish process. However, 'values' might include fairness to other societies. This chapter has argued that fairness plays a minor role. However, if it were clearly demonstrable that one country caused serious harm to another, the offending public might call for regulation on the basis that such behaviour is not 'right'. Willingness to pay would be higher and the outcome might be more regulation and international agreements that are 'fair'. While this is an interesting scenario, it is unlikely that such a burden of proof will be met in the foreseeable future for acute damages caused by global warming (damages to low-lying nations from sea level rise might be an exception, many decades in the future). Moreover, in many instances serious and demonstrable harm to others has not resulted in any significant higher willingness to pay. Examples include air and water pollution from the Soviet bloc that caused (and is still causing) environmental damage in the West. After the end of the Cold War and the collapse of the Soviet bloc, some former centrally planned countries in Eastern Europe began to do more to regulate pollution, including pollution that flows across borders to the West. However, much of that action reflects a more selfish calculation: many of these countries want to join Western institutions (notably the EU), and doing so first requires meeting Western environmental standards. In Russia, the willingness to pay to control pollution from the notorious Kola nickel smelters has increased, but that reflects not only internal fear of pollution but also, mainly, the many Western offers to pay much of the abatement costs.[16] With democratization, the public in the former centrally planned countries is now demanding that their govern-

---

15 For simplicity, this chapter does not employ the cumbersome changes in legal language in discussing these projects. Formally, per the 1995 Berlin Mandate, the projects underway are pilot activities implemented jointly projects and are intended to build experience but not to earn credits for greenhouse gas reductions. Eventually some or all of these projects will probably become part of the clean development mechanism (CDM) and thus will earn credits that can be used to offset emissions commitments in industrialized (Annex I/Annex B) countries. But rules for the CDM have not yet been agreed upon. Nonetheless, a few trades of 'credits' have taken place. In the present period of high uncertainty, for simplicity this chapter labels all of this 'joint implementation', although the very important details and terminology are still in flux.
16 Moreover, it is difficult to assess the real willingness to pay because, despite a decade of Western-led efforts to arrange financing for the smelters, there has been essentially no practical action. The latest Norwegian-led effort collapsed in November 1996. For a review from the perspective of Russia and the Kola nickel firm, see Kotov and Nikitina (1996).

ments do more to protect the environment. However, most (perhaps all) of such pressure reflects concern about environmental damage within the borders. Luckily for downwind and downstream states, some actions, such as limiting local concentrations of sulphur dioxide in the air or excess nutrients in the water, also have the incidental benefit of limiting international problems such as acid rain and Baltic Sea pollution.

Willingness to be fair may increase when societies are aware of their harms and are able to press their governments to correct them (that is, in democracies where the density of useful information is high). This proposition would be interesting to evaluate but has not been critiqued extensively. This author suspects that the role of international fairness is higher in rich democracies – witness, for example, foreign aid paid by rich democracies to developing countries and the trend to reduce the 'strings' attached to such aid. But in many countries aid budgets have been cut, and the rhetoric of foreign aid is dominated by self-interest. As rich and poor societies rely more upon markets to allocate resources, the role of self-interest (by definition) rises, which could make it increasingly difficult to achieve fairness, especially at the international level where a cohesive society does not exist.

Secondly, this chapter has not addressed the role of rhetoric. In negotiating and implementing climate agreements, fairness is often cited in support of particular policy proposals, such as the need to limit or avoid imposing regulatory obligations upon developing countries. Rhetoric certainly affects tactics, but it may not be crucial in determining the significant elements of international agreements, which each nation weighs according to self-interest. Extensive rhetoric about the need for fairness is not evidence that fairness has much influence on outcomes, and it may be evidence only that fairness is symbolically important or when international agreements have little effect on behaviour.

Thirdly, this chapter's conclusion that fairness is not very important is based on the assumption that policy actions to slow global warming will change only marginally the policies of states and the functioning of the world economy. This conclusion may not hold if a crash programme to achieve a deep cut in global carbon emissions is pursued – say, cutting carbon by 50 per cent or more worldwide over the next few decades. In that case, the agreement adopted might reflect fairness concerns more prominently because the change in policy and the need for global participation would be so radical that they would force some kind of global compact. In other words, such a change would quickly require humanity to think like a society of people, not like a collection of individual states and societies. Moreover, emissions trading would surely be on the agenda, and thus so would the formula for allocating permits. This author doubts that such radical thinking and action will happen, unless society becomes seized by fears of climatic catastrophe, which will increase the willingness to pay for abatement of greenhouse gases, geoengineering and adaptation.

REFERENCES

Banuri, T, Göran-Mäler, K, Grubb, M, Jacobson, H K and Yamin, F, 1995, Equity and social considerations, in: J P Bruce, H Lee and E F Haites, eds, *Climate Change 1995: Economic and Social Dimensions of Climate Change*, Cambridge University Press, Cambridge

The Economist, Europe's Union: Who pays for it?, 23 November 1996, pp 39–40

Hjorth, R, 1998, Implementation of Baltic Sea pollution commitments in Poland: A review of the literature, in: D G Victor, K Raustiala and E B Skolnikoff, eds, *The Implementation and Effectiveness of International Environmental Commitments: Theory and Practice*, MIT Press, Cambridge

Kotov, V and Nikitina, E, 1996, Russia wrestles with an old polluter, *Environment*, 38(9):6–11 and 32–37

Levy, M A, 1995, International cooperation to combat acid rain, in: H O Bergesen and G Parmann, eds, *Green Globe Yearbook*, 1995, Oxford University Press, Oxford

Raustiala, K and Victor, D G, 1998, Conclusions, in: D G Victor, K Raustiala and E B Skolnikoff, eds, *The Implementation and Effectiveness of International Environmental Commitments: Theory and Practice*, MIT Press, Cambridge

Roginko, A, 1998, Domestic implementation of Baltic Sea pollution commitments in Russia and the Baltic states, in: D G Victor, K Raustiala and E B Skolnikoff, eds, *The Implementation and Effectiveness of International Environmental Commitments: Theory and Practice*, MIT Press, Cambridge

Slaughter, A-M, 1995, International law in a world of liberal states, *European Journal of International Law* 6:503–538

Tol, R S J, Fankhauser, S and Pearce, D W, 1999, Empirical and ethical arguments in climate change impact valuation and aggregation (Chapter 4 of this book)

Victor, D G and Salt, J E, 1995a, Keeping the climate treaty relevant – an elaboration, circulated at the First Conference of the Parties to the Framework Convention on Climate Change, 3 April

Victor, D G and Salt, J E, 1995b, Keeping the climate treaty relevant, *Nature* 373:280–282

Victor, D G, Raustiala, K and Skolnikoff, E B, eds, 1998, *The Implementation and Effectiveness of International Environmental Commitments: Theory and Practice*, MIT Press, Cambridge

Addresses of Lead Contributors

H Asbjørn Aaheim
CICERO
University of Oslo
PO Box 1129
N–0317 Oslo
Norway

Frank Biermann
Secretariat of the German Advisory
 Council on Global Change
Postfach 120161
D–27515 Bremerhaven
Germany

Carsten Helm
Potsdam Institute for Climate Impact
 Research (PIK)
Pf 601203
D–14412 Potsdam
Germany

Juliane Kokott
Faculty of Law
University of Düsseldorf
Universitätsstraße 1
D–40225 Düsseldorf
Germany

Joanne Linnerooth-Bayer
International Institute for Applied
 Systems Analysis
A–2361 Laxenburg
Austria

Volker Linneweber
Institute for Psychology
Otto-von-Guericke University
Postfach 4120
D–39016 Magdeburg
Germany

Shuzo Nishioka
Global Environment Research Group
National Institute for Environmental
 Studies
16–2 Onogawa, Tsukuba, Ibaraki 305
Japan

Steve Rayner
Battelle Pacific Northwest
 Laboratories
901 D Street SW #900
Washington, DC 20024
US

PR Shukla
Indian Institute of Management
Vastrapur
Ahmedabad 380015
India

Richard SJ Tol
Institute for Environmental Studies
Vrije Universiteit
De Boelelaan 1115
NL–1081 HV Amsterdam
The Netherlands

Ferenc L Tóth
Potsdam Institute for Climate Impact
 Research (PIK)
Pf 601203
D–14412 Potsdam
Germany

David Victor
Council on Foreign Relations
58 E 68th Street
New York, NY 10021
US

INDEX

abatement costs 83, 84–5, 86, 91, 95, 96, 100–4, 108
ability to pay concept 7, 83, 101
acceptability 88
accessibility 142
acquired rights 187
active filter 119, 120
aggregate willingness 60
aggression 113
allocational issues 21–6
altruism 113
ambiguity 118, 122
Ambonese 136–7
Antarctic Treaty 162
Argentina 203
Aristotle 21, 50, 83, 174
Asia-Pacific region 6, 133, 134–9, 143
Association of South-East Asian Nations (ASEAN) 138
asymmetry 146–7
Australia 96, 98, 108, 202
Austria 50, 52, 56
availability heuristic 121
averaging 72–3

background allocation 20
Balinese 137
Baltic Sea 196–7
Bangladesh 137
bargaining theory 146
basic needs emissions 28
behavioural paradigm 49
benefit transfer 69–70, 72
Benthame's principle 81
Berlin mandate 187, 202
Bernoulli-Nash function 71, 75
bias 112–32
bottom-up studies 96–7, 141
burden allocation 80–93
burden sharing 4, 94–111, 123, 184, 197
burden size 147

Canada 96
capabilities 165
carbon tax 22
catastrophist view 18

CFCs, smuggling 7
Chernobyl 53
Chile 203
China 59, 135–6, 203
Coase theorem 147–8
cognitive distortion 121
collective responsibility 74
common concern of humankind 7, 162, 168–70, 181
communitarians 52
compensation 55–6, 69, 73–5
consequential equity 2, 149–50
contractarianism 21, 24, 25, 27
contracts 30, 31
contra legem 174, 175, 176, 178
contribution 59
convergence criteria 154–5
convergence framework 153–4
coral reefs 141
cornucopian view 18
cost allocation 45
cost-benefit analysis 35, 141–2
cost effectiveness 95, 96–100, 106, 147
cost efficiency 95–6, 100–4
cost function 1
cultural bias 51
cultural groups 51–2
cultural theory 46, 51, 58
culture 5, 6
customary law 161, 183

damage costs 66–7, 101–2, 104, 105
damage estimates 4, 74, 76
damage function 1
decision-making tools 141
deep ecologists 23
democracy 52, 198–9, 204–5
Department of Environment and Natural Resources 183–4
descriptive paradigm 12–13, 14–15, 16
developing countries 6, 8, 20, 145–59
 abatement costs 83
 burden allocation 90–2
 equity principle 50
 foreign aid 47, 49, 53
 joint implementation 57
 Kyoto 44, 202–4

legal rights 167
new international economic order 177–8
parity 23, 84
per capita parity 22
research access 142
role 124–5
utilitarian argument 54
vertical principle 108
vulnerability 66
WGII 60
dictatorship 105
difference principle 82
differentiated obligations 58–9
differentiation 164, 165, 169, 199–201
discount rates 32, 33
distribution equity 29
distribution principles 21–6, 82
distributive justice 46, 58, 82, 175, 177, 181
divergence of perspective 125
dvidual 36
dyads 113–14

Earth in the Balance 55
earthrights model 23
ecoconsciousness 135–9, 143
ecological farming 135–6
economics 3–5, 80–93, 94–111, 147
efficiency 4, 23, 87, 101
   developing countries 145–59
   equity dichotomy 3
   integrated assessment 11
   utilitarianism 16
egalitarianism 21, 22, 26, 36, 51, 54–5, 56–7, 106–8
   discount rates 33
   equality 58
   explicit consent 27, 28
   historical 25
   solidarity 31
emission rights 21, 22, 23, 24, 58, 84
emission trading 44, 57
endowment 59
Energy Technology Systems Analysis Programme (ETSAP) 97
environmental solidarity principle 166
environmental space 23
environment concept 116–18
environment phenomenon 118–21
equality 58, 86
equal participation principle 167, 169
equal per capita 157
equitable use principle 184
equity principle 50, 83–4
equity weighting 70–1
erga omnes 179, 180, 187
European Union 47, 197–8

ex aequo et bono 176
explicit consent 27, 28

fair share guaranteed criterion 88, 90
fatalism 51
flat rate reductions 96, 105
flexibility 99–100, 108, 147
floods 137
formal justice 162–3
France 180
freshwater protection 180–1
functionalist theories 29

GCAM 38
GDP 24, 25, 66–7, 98
Gemeinschaft 30, 31
General Agreement on Tariffs and Trade (GATT) 170, 177
Germany 180
Gesellschaft 30, 31
global circulation models (GCMs) 117, 122
global commons 115–16
Global Environmental Facility (GEF) 140, 167, 200
globalization 139, 143
global management 52
global welfare 104–8
global welfare distribution 4
grandfathered criterion 153, 157
Greece 197
Group of 77 59, 164, 203

hazardous waste disposal 3, 45–64
Helsinki Rules 180
hierarchical determinants of welfare 52
hierarchical organization 51
hierarchies 30, 31, 32–3, 49
historical obligation 28
historical responsibility 73–5
holistic perspective 12, 54, 83
holistic values 139, 142, 143
horizontal principle 106–8
hot air 57
hypothetical consent 27, 28

Iceland 96, 108, 164
illusions 120
impact valuation 65–79
implicit consent 27
impossibility theorem 104
income-based entry 156
income effect 154
income elasticity 70
income scaling 69–70
India 57, 141, 157
indigenous knowledge 135–6
indirectness 122–3

individualism 51, 53, 55, 59–60
individual rationality 88
industrialized countries 6, 8, 20
  abatement costs 83
  actors 124–5
  burden allocation 90–2
  deep ecologists 23
  equity principle 50–1
  income-based entry 156
  joint implementation 57
  Kyoto conference 44, 202–4
  new international economic order 177–8
  parity 84
  per capita parity 22
  utilitarian argument 54
  vertical principle 108
  WGII 60
inequality attraction 76
ingroup favouritism 114
institutional solidarity 36
instrumentalist theories 29
integrated assessment models (IAMs) 1, 4, 9, 141, 193
integrated assessments (IAs) 3, 11–43
interdependence 118–21
intergenerational equity 2, 32–3, 125, 182–4, 185
Intergovernmental Panel on Climate Change (IPCC) 157, 185–6
  cost allocation 45
  equity approaches 81
  integrated assessment models 141
  parity 83
  per capita emissions 150, 151–2
  Second Assessment Report 1–2, 4, 60, 62, 65, 141–2
  Special Report of Regional Impacts 142
intergroup relations 114–15
International Court of Justice 161, 174, 176, 182–3
international law 6–8, 160–72, 173–92
International Law Association 177, 180
interpersonal comparisons 105
interpretive paradigm 12–13, 14, 16, 17, 18–19
intertemporal equity trading 86
intragenerational equity 2
intra legem 162, 174–5, 176
inverse equity weighting 75–6
ius cogens 161, 179

Japan 47, 96, 97
justice 5, 6–7, 54, 104–8, 160–72, 175
  developing countries 145–59
  local 46, 85
  principles 115

Kant, I. 12, 17, 106
Korea 138, 203
Kuwait 164
Kuznet curve 155
Kyoto conference 20, 44–5
  cost effectiveness 96
  emission trading 57
  North-South differentiation 164
  US 59
Kyoto Protocol 54, 58, 108, 173, 186, 193, 194, 197, 201–4
  ratification debate 54
  sink 142

law 6–8, 160–72, 173–92
legitimacy 175
libertarianism 21, 25, 53
local concerns 133–44
local environment management 3
local impact 6
local justice 46, 85
local knowledge 141
long-range transmission of air pollutants (LRTAP) 200
loss-spreading 25

macroeconomic models 96, 98–9
mapping of discourse and human values 3, 34–7, 39
marginal abatement costs 6, 83, 91, 96, 100, 105
market efficiency 147, 148
marketing value 141–2
markets 30, 31
market solidarity 35
market utilitarianism 49, 51, 52, 53–4, 60–1
maximum value 73
mechanical solidarity 30, 31
merit 59
mesosocial systems 114
methodological individualist theories 29
Mexico 203
migration 140–1
mitigation 140
  burden 6, 157
  cost 147, 148
monetization 65
monotonicity 89, 90
mono-value 142
Montreal Protocol 7, 163–4, 166, 167, 169, 185, 196, 199, 200
moral outrage 50
mortality risks 71, 72, 73
multichannel relationships 126
multifunctionalism 30, 31
multilateral environmental agreements 179–82

Multilateral Fund 196
multivalue 142
myths of nature 18

naive models 122
NATO 199
natural debt 28
natural law 174, 175
natural sciences 14, 122
needs 59
neem tree 141
net costs 101, 106
Netherlands 97, 180, 197
New International Economic Order (NIEO) 177–8
New Zealand 59, 202
no envy criterion 87
non-governmental organizations (NGOs) 141, 199
non-institutional perspectives 29
non-marketing value 141–2
no regrets strategy 33
Norway 96, 98, 108

obligations allocation, bias 112–32
OECD 26, 74, 203
opinio iuris 183
OPOSA 183–4
organic solidarity 30
organic specialization 31
organizations 114
outgroup discrimination 114
ozone regime 163–4, 166

Pakistan 138
Pareto criterion 48–9
Pareto efficient allocation 83
Pareto improvement 146
Pareto optimality 25, 36, 87, 94
Pareto proposition 49–50
Pareto rationality 60
Pareto superior policies 57
parity 21, 22–3, 24, 28, 36, 83, 84
patents 21, 141
per capita approach 188
per capita emissions 57, 150–2, 154
per capita entitlements 22, 28, 58, 86, 152
personal families 30
personal freedom 31
perspective-specific perception 119
per unit values 70, 71–6
Philippines 183–4
Poland 197
political science 6–8
polluter pays principle 26, 73–5, 94, 96, 180, 187
population growth 18–19

population monotonicity 89
positional authority 31
positional families 30
positive illusions 120
praeter legem 174, 175, 176
preference 59–61
Price-Anderson Act 25
pricing 18–19
priority 21, 23–4, 36
procedural equity 2, 20–1, 27, 29, 149–50, 167
profligacy 18–19
property rights 18, 58, 68
proportionality 21, 25, 36, 83
prospect theory 68
protest bids 68
psychology 112–32
public information 39

Rawls, J. 21, 54, 82, 106, 145, 148–9, 162–3, 195
realism 23
reasonable use 117
reciprocity 126
reduction of complexity 121
reflexivity 16
representativeness heuristic 121, 123
resource monotonicity 89
responsibility 50, 52, 53, 54, 55, 73–5, 165
revealed preference approach 27
Rhinesalt Convention 180
rights 59–61
rights-based approaches 3, 17–18
rights of man 17, 165
right to development 165
risk analysis 11–12
Russia 57, 108, 164, 197, 202, 204

sashi 136–7
scientific knowledge 142
sea level rise 139, 140, 141
Second Assessment Report (SAR) 1–2, 60, 62, 65, 141–2
self-interest 198
self-monitoring 118–19
self-sufficiency 136
Seoul Declaration 177
Siberut Island 136
Sierra Club 55
Singapore 138–9, 143
sink capacity 83
social costs 4, 65
social identity theory 114
social judgement research 112–32
social organization 51–5
social sciences 3, 5–6, 12–19
social solidarity 51

social trap 115
social units 125–6
social vagueness 122
social welfare function 81–2, 83
societal good 16
solidarity 30–3, 36, 169
South Korea 164
sovereignty 105–6, 108, 169
specialization 30
Special Report of Regional Impacts 142
Sri Lanka 136
stand alone criterion 89–90
status 30
stewardship 52
Stockholm Declaration on the Human Environment 53, 165, 183
S trajectories 152
stream model 29
sulphur dioxide tradeable permits 57
sulphur emissions 200
sustainability 143–4
sustainable agriculture 135–6
sustainable development 181
Switzerland 50, 180

TARGETS model 38
taro pits 136
technology 141
Thailand 137
top-down studies 96
total cost minimization 23
tragedy of the commons 115
transborder damages 53
transboundary pollution 187
transfers 84–5, 86, 87
treaties 161
tropical forests 142
trusteeship approach 24
Turkey 197
Tuvalu 136, 140–1, 142

Ukraine 108, 164
ULYSSES Project 39
umbrella group countries 57
unanimity 87
United National Convention on the Law of the Non-Navigational Uses of International Watercourses 180
United Nations Charter 106
United Nations Charter on Economic Rights and Duties of States 177
United Nations Conference on Environment and Development (UNCED) 181–2
United Nations Convention on Biological Diversity (UNCBD) 7, 162, 164, 166, 181
United Nations Convention on the Law of the Sea (UNCLOS) 162, 199
United Nations Declaration of the General Assembly 177
United Nations Development Programme (UNDP) 167
United Nations Environment Programme (UNEP) 167
United Nations Framework Convention on Climate Change (UNFCCC) 20, 80, 162, 173, 181–2, 185–7, 188–9, 193, 202–3
  ability to pay 83
  burden size 147
  common concern 6–7
  differentiated obligations 58–9
  equity 149
  intergenerational equity 185
  mitigation 157
  past activities 165
  solidarity 166
  Tuvalu 140
  vertical principle 106
United Nations Watercourses Convention 180
United States 47, 202
  damage costs 66
  emissions trading 57
  Kyoto conference 54, 59
  Multilateral Fund 196
  reduction costs 97
  waste export 52–3
utilitarianism 3, 16–18, 25, 47–9, 52–4, 81–2
utility function 71
utility maximizing theories 29

valuation 68–71, 76–7
value of a statistical life (VOSL) 69, 70, 71–2, 73
vertical principle 106–8
Vienna Convention on the Protection of the Ozone Layer 185
voluntary market approaches 55
vulnerability 66, 67, 139

weighted votes 167
welfare analysis 96
welfare economics 4, 35–6, 48, 50, 60, 70
welfare function 5, 70, 71, 75
willingness to accept (WTA) 4, 66, 68–72, 76
willingness to pay (WTP) 4, 8, 60, 66, 68–72, 76, 101, 195, 196–8, 202, 203, 204
World Bank 49, 60, 167
World Trade Organization 177
WRE trajectories 152